THE END OF AN ERA IN SPACE EXPLORATION

From International Rivalry to International Cooperation

SERIES EDITOR
Dr. Horace Jacobs — *Lockheed-California Company*

AUTHOR
Dr. J. C. D. Blaine — *Professor Emeritus, Department of Economics, University of North Carolina*

ART STAFF
Larry Chan — *Lockheed-California Company*

Thanks are due Sarah Mason, University of North Carolina at Chapel Hill, for typing the manuscript and Diana Law for final preparation of the manuscript for printing.

Front Cover and Title Sheet: Courtesy of Lockheed-California Company

Front Cover Illustration: Apollo Soyuz Rendezvous, courtesy of the National Aeronautics and Space Administration

Frontispiece: First Viking Orbiter I photo of Phobos, a satellite of Mars. Taken July 25, 1976. Courtesy of the National Aeronautics and Space Administration.

Dedicated to:

All those who have contributed and are contributing to progress in space exploration and travel.

We choose to go to the moon in this decade and do other things not because they are easy, but because they are hard, because that goal will serve to organize and measure the best of our energy and skills.

John F. Kennedy
President of the United States of America

May, 1961

AN AMERICAN ASTRONAUTICAL SOCIETY PUBLICATION

THE END OF AN ERA IN SPACE EXPLORATION

from International Rivalry to International Cooperation

by J.C.D. Blaine

Distributed by UNIVELT, INC., P.O. Box 28130, San Diego, CA 92128

AAS Publications Office
P. O. Box 28130
San Diego, California 92128

Affiliated with the American Association for the Advancement of Science
Member of the International Astronautical Federation

ISBN 87703-080-4

Printed and Bound in the U.S.A.

NATIONAL AIR AND SPACE MUSEUM SMITHSONIAN INSTITUTION

WASHINGTON, D. C. 20560

July 13, 1976

Dr. Horace Jacobs
American Astronautical Society
Publications Office
P.O. Box 746
Tarzana, CA 91356

Dear Dr. Jacobs:

Thank you for your letter of 1 July enclosing the page proofs of Professor Blaine's book. I have perused it with interest and enjoyed his organization and treatment, concentrating, as the author does, on the comparative evolution of United States and Soviet Union programs. It is a most readable summary of the incredible space flight accomplishments of less than twenty years.

I hope you will be able to publish Professor Blaine's work this year and look forward to seeing the final product.

Sincerely,

F. C. Durant, III
Assistant Director
Astronautics

cc: P. H. Bolger

ABOUT THE AUTHOR

Dr. J. C. D. Blaine, Professor of Economics, Emeritus, University of North Carolina, Chapel Hill, North Carolina, has had a distinguished career as professor of economics (transportation) at the University of North Carolina at Chapel Hill. Special assignments have included: Participant in the Southeast American Assembly on the *Future of American Transportation,* University of Miami, 1972; Consultant to Director General, National Council of Applied Economic Research, New Delhi, India, *Regional Transport Survey of Jammuand Kashmir,* 1971; *Air Transportation Survey,* 1963; Transportation Consultant, The Research Triangle in Tehran, Iran, 1968; also conducted Independent Study of Transportation Relating to Western Europe and the European Economic Community, 1963, and *Survey of Potential Shipping in the Caribbean,* 1956. Served as visiting Professor of Economics, University of Utah, 1959, Acting Head, Department of Economics, University of New Brunswick, 1945-46, and Captain Canadian Army 1941-45. Professor Blaine is the recipient of many honors and the author of numerous articles and reports. He has taken an active interest in air and space transportation.

AUTHOR'S FOREWORD

This historical treatise is a non-technical presentation of the events and achievements of the first era of the Space Age characterized by great advances in technology engendered in part by a contest between the Soviet Union and the United States in the exploration of the moon and planets of the solar system. The Soviet Union has relied heavily on the use of unmanned spacecraft and electronic devices, shrouded in secrecy as to information pertaining to space missions. The United States, in contrast, has relied more heavily on the use of manned spacecraft and electronic devices, under a policy of openness and disclosure of information pertaining to space missions.

Both nations succeeded in achieving spectacular firsts in their spatial ventures. The United States was first to land men on the surface of the moon and to carry out extensive on-the-surface scientific and exploratory extravehicular excursions. The Soviet Union was first to soft-land an automatic spacecraft on the surface of the moon which, after gathering soil samples, returned to earth. Spectacular feats have been achieved by both nations in the exploration of the moon and planets of the solar system.

Attention is being focused on the exploration of the moon and planets of the solar system and upon distant stars in interstellar space. There are places in space to which man may never go and he must rely on the transmission of electrical signals for contacts with them. There are other places in space to which he may ultimately go after extensive unmanned explorations. The investigation of space is becoming more extensive and as a result more costly and time consuming.

Emphasis is shifting from the exploration to the exploitation of space with increased attention being given to an integrated space transportation network, space stations, and shuttle spacecraft. In addition, considerable thought is being given to the establishment of earth-orbiting production centers for the manufacture of purer products, under conditions of weightlessness, than can be produced under earth-gravitational conditions.

If the potentials of space are to be achieved more fully for the benefit of mankind, greater cooperation and participation will be necessary. International efforts should involve the free interchange of scientific and technological information, as well as joint projects for the accumulation, analysis, and application of space knowledge. There is increasing evidence of the desire and willingness on the part of the Soviet Union and the United States to cooperate and participate more fully in the use of space. It is hoped that the initial steps taken by these nations will be expanded into intensive and genuine efforts in international cooperation and participation based on mutual trust and goodwill. If this occurs, mankind may enjoy more fully the benefits of future space exploration and exploitation under more peaceful circumstances.

AMERICAN ASTRONAUTICAL SOCIETY

Publications Office, Post Office Box 28130, San Diego, California 92128

Science and Technology

Vol. 1 Manned Space Reliability Symposium, 1965 (112 pages)—$15.00
Vol. 2 Towards Deeper Space Penetration, 1965 (182 pages)—$15.00
Vol. 3 Orbital Hodograph Analysis, S. P. Altman, 1965 (150 pages)—$15.00
Vol. 4 Scientific Experiments for Manned Orbital Flight, 1965 (372 pages)—$25.00
Vol. 5 Physiological and Performance Determinants in Manned Space Systems, 1965 (220 pages)—$15.00
Vol. 6 Space Electronics Symposium, 1965 (404 pages)—$25.00
Vol. 7 Theodore Von Karman Memorial Seminar, 1966 (140 pages)—$15.00
Vol. 8 Impact of Space Exploration on Society, 1966 (382 pages)—$25.00
Vol. 9 Recent Developments in Space Flight Mechanics, 1966 (280 pages)—$20.00
Vol. 10 Space Age in Fiscal Year 2001, 1967 (458 pages)—$30.00
Vol. 11 Space Flight Mechanics Symposium, 1967 (618 pages)—$35.00; Microfiche Suppl.—$6.00 extra
Vol. 12 The Management of Aerospace Programs, 1967 (392 pages)—$20.00
Vol. 13 The Physics of the Moon, 1967 (260 pages)—$20.00
Vol. 14 Interpretation of Lunar Probe Data, 1967 (270 pages)—$20.00
Vol. 15 Future Space Program and Impact on Range and Network Development, 1967 (583 pages)—$30.00
Vol. 16 The Voyage to the Planets, 1968 (184 pages)—$15.00
Vol. 17 Use of Space Systems for Planetary Geology and Geophysics, 1968 (623 pages)—$35.00; Microfiche Suppl.—$4.00 extra
Vol. 18 Technology and Social Progress, 1969 (170 pages)—$15.00
Vol. 19 Exobiology—The Search for Extraterrestrial Life, 1969 (184 pages)—$15.00
Vol. 20 Bioengineering and Cabin Ecology, 1969 (162 pages)—$10.00
Vol. 21 Reducing the Cost of Space Transportation, 1969 (264 pages)—$20.00
Vol. 22 Planning Challenges of the 70's in the Public Domain, 1970 (504 pages)—$30.00; Microfiche Suppl.—$12.00 extra
Vol. 23 Space Technology and Earth Problems, 1970 (418 pages)—$25.00; Microfiche Suppl.—$12.00 extra
Vol. 24 Aerospace Research and Development, 1970 (500 pages)—$30.00
Vol. 25 Geological Problems in Lunar and Planetary Research, 1971 (750 pages)—$40.00
Vol. 26 Technology Utilization Ideas for the 70's and Beyond, 1971 (312 pages)—$25.00
Vol. 27 International Cooperation in Space Operations and Exploration, 1971 (194 pages)—$16.00
Vol. 28 Astronomy from a Space Platform, 1972 (416 pages)—$30.00
Vol. 29 Space Technology Transfer to Community and Industry, 1972 (196 pages [Microfiche only])—$10.00
Vol. 30 Space Shuttle Payloads, 1973 (532 pages)—$35.00
Vol. 31 The Second Fifteen Years in Space, 1973 (212 pages)—$20.00
Vol. 32 Health Care Systems, 1974 (264 pages)—$20.00
Vol. 33 Orbital International Laboratory, Third & Fourth Oil Symposia, 1974 (322 pages)—$25.00
Vol. 34 Management and Design of Long-Life Systems, 1974 (198 pages)—$16.00
Vol. 35 Energy Delta, Supply vs. Demand, 1975 (604 pages) —Paperback—$40.00; Microfiche—$25.00
Vol. 36 Skylab and Pioneer Report, 1975 (160 pages)—$16.00
Vol. 37 Space Rescue & Safety, 1974 (294 pages)—$25.00
Vol. 38 Skylab Science Experiments, 1975 (274 pages)—$25.00
Vol. 39 Environmental Control and Agri-Technology, 1976 (346 pages)—$30.00
Vol. 40 Future Space Activities, 1976 (182 pages)—$20.00 (incl. numerical & author index for all Goddard Memorial Symposia)
Vol. 41 Space Rescue & Safety 1975, (230 pages)—$20.00

Advances in the Astronautical Sciences

Vol. 1-5 AAS Proceedings 1957-60. Available in Microfiche. $20.00 per volume.
Vol. 6 Sixth Annual Meeting Proceedings, 1961 (968 pages)—$35.00
Vol. 7-8 Third West Coast & Seventh Annual Meeting Proceedings, 1961-63. Available in Microfiche. $20.00 per volume.
Vol. 9 Fourth Western Meeting Proceedings, 1963 (910 pages)—$35.00
Vol. 10 Manned Lunar Flight, 1963 (310 pages)—$15.00
Vol. 11 Eighth Annual Meeting Proceedings, 1963 (808 pages)—$35.00
Vol. 12 Scientific Satellites, 1963 (262 pages)—$15.00
Vol. 13 Ninth Annual Meeting: Interplanetary Missions, 1963 (690 pages)—$30.00
Vol. 14 Physical and Biological Phenomena in a Weightless State, 1963 (382 pages)—$20.00
Vol. 15 Exploration of Mars, 1963 (634 pages)—$35.00
Vol. 16 Space Rendezvous, Rescue, and Recovery, 1963; Part I (1028 pages)—$35.00; Part II (380 pages)—$20.00
Vol. 17 Bioastronautics—Fundamental and Practical Problems, 1964 (128 pages)—$15.00
Vol. 18 Lunar Flight Programs, 1964 (630 pages)—$35.00
Vol. 19 Unmanned Exploration of the Solar System, 1965 (1000 pages)—$45.00
Vol. 20 Post Apollo Space Exploration, 1966; Part I (572 pages)—$35.00; Part II (648 pages)—$35.00
Vol. 21 Practical Space Applications, 1967 (508 pages)—$30.00
Vol. 22 The Search for Extraterrestrial Life, 1967 (388 pages)—$25.00; Microfiche Suppl.—$3.00 extra
Vol. 23 Commercial Utilization of Space, 1968 (512 pages, plus 24 microfiches)—$35.00
Vol. 24 Exploration of Space, 1968 (363 pages)—$20.00
Vol. 25 Advanced Space Experiments, 1969 (530 pages)—$35.00
Vol. 26 Planning Challenges of the 70's in Space, 1970 (470 pages)—$30.00; Microfiche Suppl.—$10.00 extra
Vol. 27 Space Stations, 1970 (606 pages)—$35.00
Vol. 28 Space Shuttles and Interplanetary Missions, 1970 (488 pages)—$30.00
Vol. 29 The Outer Solar System, 1971; Part I (618 pages)—$35.00; Part II (740 pages)—$35.00
Vol. 30 International Congress of Space Benefits, 1974 (528 pages)—$35.00
Vol. 31 The Skylab Results, 1975 (1174 pages, Microfiche only)—$45.00
Vol. 32 Space Shuttle Missions of the 80's, 1976 (approx. 1000 pages)—$70.00; Microfiche Suppl.—$60.00
Vol. 33 AAS/AIAA Astrodynamics Conference 1975, (approx. 500 pages)—$35.00; 1976 Microfiche Suppl.—$35.00

Special Volumes

1. Weightlessness—Physical Phenomena and Biological Effects, 1961 (182 pages)—$15.00
2. Lunar Exploration and Spacecraft Systems, 1962 (214 pages)—$15.00

DISTRIBUTED BY **UNIVELT, INC., P.O. Box 28130, San Diego, California 92128**

PREFACE

This historical presentation of space exploration activities of the United States and the Soviet Union encompasses only a small part of their extensive space programs. It does not deal with space activities concerned primarily with scientific investigation of space phenomena, meteorology, communication, and economic appraisals of the world's human and material resources. It is basically a coverage of unmanned and manned moon- and solar planet- probing space flights as a vehicle for understanding the need for increased international cooperation in the exploration and use of space.

It has been a most challenging and rewarding undertaking for the author, especially in attempting to recreate the trying and thrilling space flights of the astronauts and cosmonauts who, as true and courageous pioneers, ventured into the yet unknown but unfolding realm of space. This has been difficult because of the nature of the sources of information which at times were fragmentary, condensed, and restricted.

The author expresses his thanks to: Dr. Horace Jacobs, Series Editor of the American Astronautical Society, and his staff for their interest and efforts in editing and preparing the manuscript for publication; Mrs. Robert H. Goddard for use of the photograph of the first liquid-fuel rocket to be flown with her late husband standing beside it; Mr. Les Gaver, Chief, Audio-Visual Branch, National Aeronautics and Space Administration, for the NASA photographs used as illustrations throughout the book, and Mrs. Sarah Mason, Secretary, Department of Economics, the University of North Carolina at Chapel Hill, for typing the manuscript. Special thanks go to my wife Eva for her assistance in reading the galley proofs for the book.

Chapel Hill, North Carolina. J.C.D.B.

AMERICAN ASTRONAUTICAL SOCIETY

Purpose and Scope

The American Astronautical Society is the singular professional, scientific and engineering organization in the United States dedicated solely to the advancement of the astronautical sciences.

The basic aim of the Society is to further the progress of space exploration and exploitation by encouraging the exchange of ideas and information among scientists and engineers. To promote this aim, the Society strives to stimulate interest and development of the astronautical sciences through publications, special symposia, and local and national meetings.

The Society purposes to present those facts and theories which will further advance the astronautical sciences and space technology, guide the nation's astronautical programs toward maximum achievement in space exploration, and inspire young scientists and engineers to further these causes.

Quality, rather than quantity, is emphasized in society membership. Professional men desiring to participate in stimulating technical meetings and activities on local and national levels are invited to apply.

Membership Information

The Society is composed of six individual types of membership. The grades of membership and the requirements for these grades are:

FELLOWS shall be persons who have made a direct and significant contribution to the field of astronautics. Candidates must possess all of the qualifications of the Senior Member, and must in addition have made significant contributions in one of the astronautical sciences, and have done original work of direct value to astronautics. The Board of Directors and Fellows shall elect all applicants for this grade after consideration of the eligibility of the applicant.

SENIOR MEMBERS shall be persons who have been engaged in professional work in any of the scientific fields related to astronautics. An applicant for this grade shall have had not less than ten years of experience in one of the astronautical sciences, and shall have acquired a recognized standing in that field. The professional activity of the applicant shall be considered to be a direct contribution to astronautics. Applicants for this grade shall be recommended by one or more persons of whom one shall be a Fellow or Senior Member of the Society.

MEMBERS shall have an active interest in astronautics and shall have had six years of experience and training in some branch of science or related work which contributes toward the advancement of astronautics. Applicants for this grade shall be recommended by one or more persons who have attained the grade of Member of the Society.

STUDENT MEMBERS shall have an interest in astronautics and shall be taking a full-time course at a recognized educational institution, college or university, or shall be under 18 years of age. To qualify for this grade, applicants shall be recommended by one or more persons of Professional stature.

AFFILIATE MEMBERS shall be over 18 years of age and shall have an interest in astronautics and desire to assist in its advancement. To qualify for this grade, applicants shall be recommended by one or more persons, but preferably by a Member of the Society.

HONORARY MEMBER is a designation granted to those persons who have attained a position of distinction in the field of astronautics. The Honorary Member shall have made an outstanding contribution directly relevant to astronautics. The Board of Directors and Fellows shall elect all persons to this classification.

Benefits To Members

1. Participation in international national and regional symposia.
2. Journals and other publications published by the AAS.
3. Publication of members papers in AAS publications.
4. Assistance of the business office on special problems.
5. Employment services by exchange of information by members on a voluntary basis.
6. Affiliation with contemporaries in the astronautical sciences.

For further information regarding membership in the

AMERICAN ASTRONAUTICAL SOCIETY

Contact the business office

6060 Duke Street
Alexandria, Virginia 22304
(703) 751-7323

CONTENTS

PROLOGUE
HISTORICAL BACKGROUND

The history of modern rocketry is relatively brief but the knowledge of rocket propulsion existed centuries before the birth of Christ. There is little evidence that those who used powder-propelled rockets in the earlier stages of their development understood the scientific principles underlying their propulsion. The basis for this understanding was established by Sir Isaac Newton during the latter part of the seventeenth century when he enunciated the laws of motion.[1]

Rockets are not propelled in a forward direction by the pressure exerted upon the external air but by the pressure of escaping gas exerted upon the surface of their combustion chambers, opposite the nozzles through which the gas escapes. The action of the escaping gas upon these surfaces causes rockets to react by moving in the opposite direction or in the direction of the force exerted. Since the surfaces of the combustion chambers are part of the rockets, the rockets are thrust forward as a result of this action and reaction.

This principle may be demonstrated by allowing the air to escape suddenly from an inflated balloon and, at the same time, releasing it. The balloon will travel erratically in a direction opposite to that of the escaping air until it is deflated. Then it will cease to travel because no further pressure will be exerted upon its inner surface opposite the opening through which the air escaped. In a vacuum, a balloon would move more rapidly and further, because there would be no resistance to its motion.

Archytas of Tarentum[2] is reported to have built a wooden dove, during the fourth century B.C., driven by some manner of jet propulsion. It is also claimed that Heron (Hero),[3] a philosopher of Alexandria, constructed an engine, the Aeolipile, about 53 B.C., operated by steam jets emitted from nozzles placed tangent to the surface of a sphere. This caused it to revolve on its axis in keeping with the action-reaction principle. It is doubtful that Heron understood this scientific principle, which centuries later became known as Newton's Third Law of Motion. Little historical reference was made to rocket propulsion prior to the eleventh century, but thereafter interest in rockets propelled by blackpowder revived.

Near the beginning of the sixteenth century, a Chinese bureaucrat, named Wan-Hu, was said to have attempted a flight in a rocket machine, consisting of two kites and a seat to which were attached forty-seven rockets.[4] Some coolies lit the rockets and Wan-Hu and the rocket machine disappeared in a cloud of smoke.

During the eighteenth century, increased attention was given to rockets as weapons, following the decisive role they played in thwarting the British military campaign in India. Hydar Ali, Prince of Mysore, organized a rocket gunnery corps inflicting heavy losses upon the British at Seringapatam in 1792 and 1799.[5]

Sir William Congreve, a British colonel, contributed much to the development and application of rockets as military weapons.[6] In 1806, rockets designed by him were used in a naval attack upon Boulogne, inflicting great destruction upon that city. The following year, some twenty-five thousand rockets were launched against Copenhagen, causing its destruction by fire. In 1813, Leipzig suffered a similar fate and surrendered after three days of devasting rocket attacks.

Rockets, as weapons of war, were overshadowed for about a half-century by rifled cannons and guns which proved to be more accurate in hitting targets. Following the interlude, interest in rockets revived but it was not until near the end of World War II that it increased in momentum.

REFERENCES

1. Newton's first law deals with the uniformity of motion, holding that a body moves uniformly in a straight line undisturbed and continuously, unless it is acted upon by some external force. His second law has to do with acceleration and states that a body (mass) if acted upon by a given force moves in the direction along which the force acted. Newton's third law (here referred to as action and reaction) states that whenever a body is acted upon by another body, it exerts an equal and opposite force upon that body. (*Encyclopedia Americana,* 1964, Vol 20, p. 300)

2. *The Encyclopaedia Britannica,* Vol. I, p. 261; Vol. II, p. 446. (11th edition, 1910)

3. *Ibid.,* Vol. XIII, pp. 378-379.

4. Andrew G. Haley, *Rocketry and Space Exploration,* (New York: D. Van Nostrand Company, Inc., 1958) p. 11. (Also: Beryl Williams and Samuel Epstein cited below, pp. 56-57)

5. Willy Ley, *Rockets, Missiles and Space Travel,* (Rev. Ed.) (New York: The Viking Press, 1957), p. 67. (Also: Beryl Williams and Samuel Epstein, pp. 56-57)

6. Beryl Williams and Samuel Epstein, *The Rocket Pioneers, On the Road to Space,* (New York: Julian Messner, Inc., 1959) pp. 3-30.

Chapter I

PIONEERING STAGES OF MODERN ROCKETRY

If modern rocketry was conceived in the 1800's and born in the first quarter of the twentieth century it started to toddle in the early 1930's and by 1940 was breaking into a run.

Andrew G. Haley (1958)[1]

Three pioneers stand out in modern rocket research and experimentation during the latter part of the nineteenth and early part of the twentieth century. Like many pioneers, they were ignored largely because their thinking was far advanced for their time, but after years of rejection and frustration their contributions to the science of rocketry were acknowledged. This trio consisted of a Russian, an American and a German, each of whom was unaware of the efforts and accomplishments of the other, and arrived independently at similar conclusions.

Konstantin Eduardovich Tsiolkovskii, a Russian born on September 5, 1857, is generally recognized as the father of the science of rockets based on the principle of action and reaction.[2] This self-educated and deaf schoolmaster compiled the first truly scientific paper on rockets and space travel in 1895. The paper was submitted for publication around 1898 but was not published until 1903, when it appeared in the *Science Survey*. He lived in obscurity for years and denied himself and family much in order to satisfy his scientific and intellectual curiosity about space travel.

Earlier in life, he gave considerable thought to elongated metallic-skinned lighter-than-air ships, believing that they could be built to carry engines with sufficient power to make them dirigible. As he grew older, his interest turned to rockets and by 1920, he was giving a great deal of thought to reaction motors with sufficient power to lift a spacecraft beyond the earth's atmosphere. One of his major theoretical contributions was associated with the use of liquid fuels as rocket propellants, but he did not conduct actual experiments with such fuels nor build rocket motors or spaceships. Much of his time was devoted to the formulation of mathematical concepts and computations relating to the science of rocketry. *He is credited with being the first to conclude that liquid fuels were essential for spacecraft powered by reaction motors.* It was his contention that if men were to travel in space, it would have to be in spaceships propelled in this manner. Although he was acquainted with the principles underlying rocket propulsion and was probably the first to draft plans for a spaceship, powered by a reaction motor, he was not, however, the first to design and build a reaction engine.

Tsiolkovskii published several articles in Russian magazines and newspapers dealing with rockets and space travel as late as 1911, but most of them were ignored because they were too technical for lay readers and treated with indifference by those who should have understood them. Following World War I, he published a book about space travel entitled *Outside of the Earth*, receiving some recognition in general from lay readers but not from scientists. In 1924, the Soviet Government finally recognized him and his contributions to the science of rocketry and his treatise, published in 1903, was republished together with his other writings. On his seventy-fifth birthday, he was again honored when he received national acclaim and articles about him and his work appeared in leading Russian newspapers and magazines.

This Russian space pioneer died in 1935 when rocketry had gained the stature of a science. Throughout his life, he visualized rockets as means of travel in space not as instruments of war. It was not until October 4, 1957, that his country fulfilled in part his dream of travel in space when it placed the first unmanned satellite, Sputnik, in an orbit about the earth.

Robert Hutchings Goddard, an American scientist, was another noted pioneer of rockets and a proponent of their use for penetrating space.[3] He was born in Worcester, Massachusetts on October 5, 1882 and spent the earlier part of his life in Boston returning to Worcester in 1898. In 1904, he entered the Polytechnic Institute to study physics and became involved in experiments with static rockets, clamped in frames and charged with black powder. He graduated with a Bachelor of Science Degree in 1908 and the following year became an instructor at the Polytechnic Institute at Worcester. In 1910 and 1911, he received the Master and Doctorate degrees from Clark University, after which he went to Princeton on a one-year scholarship. Following a period of illness, he returned to Clark University in 1914 to instruct in physics and to continue his research in rocketry.

He devoted much of his efforts during the early stages of his research to experiments with solid fuels, both black and smokeless powder, as rocket propellants, and to the thermal efficiency of rockets, relating to combustion chambers and nozzles. In 1916 he prepared a report on his research for submittal to foundations but little interest was shown in it. Finally, the Smithsonian Institution granted him $5,000, enabling him to continue his rocket experimentations.

As early as 1917, Professor Goddard was focusing his attention on problems relating to the storage of liquid fuels in containers, fed separately into combustion chambers. He became involved in military rocketry during World War I and moved to Mount Wilson Observatory at Pasadena, California. In November 1918, he conducted rocket demonstrations for the Army at the Aberdeen Proving Grounds in Maryland.

He completed, by the end of May 1919, a report setting forth his findings and conclusions entitled: *A Method of Reaching Extreme Altitudes*, and submitted it to the Smithsonian Institution.[4] A section of the report presented his belief in the possibility of building a rocket capable of reaching the Moon.[5] He theorized that if it carried enough flash powder, set to explode when the rocket impacted the surface of the moon, the resulting flash would be observable from the earth with the aid of a powerful telescope.[6] The New York Times carried an editorial ridiculing the idea of a rocket crash-landing on the lunar surface,[7] and an unnamed editor criticized him severely for his flash-powder technique, although Goddard's conclusions were based upon sound theoretical analyses.[8]

Professor Goddard gave increased attention to liquid fuels possessing greater potential as rocket propellants than smokeless powder. He designed and built a high pressure pump in 1921 and the following year tested his first liquid-fuel rocket engine. Near the end of 1925, he tested his second liquid-fuel engine, which did not prove satisfactory. *The third engine was tested successfully on March 16, 1926, propelling the first rocket ever to be flown with a liquid-fuel engine.* He reported this accomplishment to the Smithsonian Institution.[9]

Dr. Goddard and the First Liquid Fuel Rocket to be Successfully Flown, March 16, 1926.

The primary objective of his rocket experiments was to discover a more effective way of obtaining meteorological information at high altitude. By 1929, he felt that he had developed a rocket suitable for this purpose and on July 17, of that year, launched an 11-foot rocket from a 60-foot tower. It rose only about ninety feet and after traveling about one hundred and seventy-one feet nosed over and crashed.[10]

Colonel Charles A. Lindbergh became interested in Dr. Goddard's experiments and was influential in persuading Daniel Guggenheim to give financial assistance for the furtherance of rocket research. Clark University received a Guggenheim Foundation Grant and he was given a two-year leave of absence to devote full-time to rocket research. He moved to Mescalero Ranch in New Mexico about three miles from Roswell to continue his research and on December 30, 1930, launched an 11-foot rocket, weighing 35 pounds without fuel, which rose to a height of 2,000 feet and attained a maximum speed of 500 miles an hour. Following this, he directed his attention to problems of stabilization and by April 1932, launched his first gyroscopically controlled rocket.

At the end of his leave of absence, Dr. Goddard returned in June to Clark University to resume his teaching duties, which were again interrupted in August 1934, when he received another Guggenheim Foundation Grant and returned to New Mexico to further his research activities. On March 8, 1935, he launched a pendulum-controlled rocket, which proved to be partially successful. After attaining an altitude of about a thousand feet the pendulum lost control causing the rocket to turn over. It then traveled two miles, reached a maximum speed of over seven hundred miles an hour, and crashed. Turning once more to the gyroscope principle of control, he launched on March 28, 1935, another rocket which rose to an altitude of 4,800 feet, turned horizontally under its guidance system, flew some 13,000 feet and attained a maximum speed of about 550 miles an hour before crashing to earth. In May of that year, Goddard launched another rocket to a height of 7,500 feet, guided by a gyroscope-controlled vane.

A report of his research activities and findings relating to liquid-fuel rockets was published by the Smithsonian Institution during 1936. This report updated his previous report giving special attention to his experiments conducted in New Mexico between July 1930 and July 1932, and between September 1934 and September 1935.[11] Upon the outbreak of World War II, he again became involved in war-time research, but did not live to see high-altitude rockets lift-off from White Sands Proving Ground near the Mescalero Ranch where he had spent so much time in rocket research. Following a throat operation, he died on August 10, 1945.

Hermann Oberth was born on June 25, 1894, at Hermannstadt, Siebenbürgen, then a part of the Austro-Hungarian Empire, now in Rumania.[12] His interest in rocketry began at an early age and proved so captivating that he rejected his father's desire for him to become a medical doctor. He entered the University of Munich as a medical student but much of his time was given to the study of mathematics and astronomy. On the outbreak of World War I, he was drafted into Austro-Hungarian military service but served for the most part in the medical corps. After the conclusion of the war, he returned to the University of Munich and later attended the University of Heidelberg, where he continued his interest in mathematics and astronomy. By 1922, he had made significant progress in the formulation of theoretical concepts relating to rocketry and space travel, most of which were conceived originally and independently, centering about passenger-carrying rockets using liquid fuel as a propellant.

He had considerable difficulty with the publication of his study, *Rakete zu den Planetenraumen,* (The Rocket into Interplanetary Space). Finally in 1923, he succeeded in having it published by paying most of the printing costs.[13] The study did not arouse much interest on the part of scientists, although considerable interest was shown by lay readers. It was based primarily on theoretical conclusions rather than empirical facts tested by experimentation, and contained numerous diagrams difficult for the layman to interpret.

Oberth became associated with Fritz Lang, the noted German movie director during the late 1920's in the production of the film *Frau im Mond* (The Girl in the Moon).[14] He acted primarily as a consultant in designing a mock spaceship and advising how to portray imaginary travel in space. He undertook to build an actual rocket to be launched at the premiere of the film but it was never completed.

Like the other pioneers in the science of rockets, Hermann Oberth was in advance of his time and did not receive the recognition he so rightly deserved until the possibility of space travel had become more widely accepted. In 1929, he was awarded the Rep Hirsch prize of 5,000 francs, the amount being doubled in recognition of his outstanding contributions to the science of rocketry. He resumed his teaching responsibilities at the Mediash School after completing his leave of absence and for several years remained in obscurity. In 1938, he was invited to participate in a rocket research program held at the Technical University of Vienna, reintroducing him to the world of rocketry science.

On the outbreak of World War II, he became involved in military rocket activities but his tour of duty proved to be frustrating in that he was still not a German citizen. He transferred to the Technical University of Dresden in 1940. That year he became a German citizen and later transferred to Peenemünde, the German rocket center, where he was assigned a minor role in the development of rockets.

The Allied Forces arrested him when they occupied Germany, but he was released later. Although, he had spent a lifetime in the study of rockets and had made outstanding contributions to the science of rocketry, he was not among the leading German rocket scientists who were brought to the United States to participate in its rocket development program. He spent some time in Italy, and then went to Nuremberg to resume teaching. In 1955, he was officially invited to come to the United States to participate in its missile development program at the Redstone Arsenal.

Space societies played an important role in keeping alive interest in rocketry during the early stages of rocket development.[15] One of these societies was the German *Verein für Raumschiffahrt,* (Society for Space Travel) founded on June 5, 1927.[16] *Its first president was Johannes Winkler, who independently constructed and launched successfully, on March 14, 1931, the first liquid-fuel rocket in Europe.* The growth of the society was rapid and within a year its membership had increased to 500, and by 1929 had grown to nearly nine hundred. It had no well-defined objectives but did build rockets that flew with some degree of success. After six years, the society was discontinued and its activities taken over by the German Military Authorities. However, it was eventually superseded by the Deutsche Gesellschaft für Raketentechnik und Raumfahrt (German Society for Technology and Space Flight) active in the field of astronautics.

The American Interplanetary Society was organized by G. Edward Pendray and David Lasser during March, 1930, but its name was changed to The American Rocket Society in 1934. The primary purpose of the society was the promotion of interest in planetary expeditions. At its inception, the society was supported by Sir Hubert Wilkins, a descendant of John Wilkins, Bishop of Chester, who had written about space travel in 1638.[17] The membership of the society totaled 6,000 by 1956, including leading scientists, engineers, and technicians in rocketry and related areas. During the early stages of its growth it built rockets, but later discontinued this activity and devoted its energies to sponsoring publications and lectures. It later merged with the Institute of the Aeronautical Sciences to form the American Institute of Aeronautics and Astronautics.

This development of the American Interplanetary Society and of the American Rocket Society was paralleled in England by the formation in 1933 of the British Interplanetary Society, which still actively pursues its interest in space flight.

The astronautical and interplanetary societies in the Western countries consider both Hermann Oberth and Robert Goddard to be their spiritual fathers. For instance, the American Astronautical Society, founded in 1953, holds an annual Goddard Memorial Meeting and Hermann Oberth, a fellow of the Society, was the recipient of the AAS Space Flight Award in 1955.

The Treaty of Versailles of 1919, stated that Germany could not rearm by means of conventional weapons, encouraging the Germans to develop rocket weaponry. By 1929, they had made significant progress in this respect under General Walter Dornberger. The German army tested two rockets near the Baltic Sea during December 1934, one of which reached an altitude of 1.4 miles. Rocket experiments were also conducted at Peenemünde, a testing site, located at the mouth of the Oder River near a small fishing village on the island of Usedom. There German rocket scientists, including Professor Wernher von Braun,[18] were making significant progress in the science of rocketry and by the autumn of 1939 launched a liquid-fuel rocket, weighing about a ton, which attained a height of five miles after a burning time of 45 seconds. Upon reaching its maximum altitude, a parachute opened allowing it to drift to a landing in the Baltic Sea from where it was recovered.

Adolph Hitler had committed Germany to war by that time and his early military successes were based on conventional weapons, causing him to disregard rockets and provide little encouragement or financial assistance to General Dornberger, The General and his associates had ready for testing the A-4 rocket missile (later known as the V-2 rocket) by October 1942, which they felt had to be launched successfully if their program at Peenemünde was to receive the recognition necessary for survival. The rocket, when launched, ascended perpendicularly, for a given period and then tilted, with an exhaust velocity of 6,500 feet per second, giving it a speed of approximately 650 miles an hour within 20 seconds. Fifty-four seconds after lift-off the reddish exhaust flame disappeared but the rocket sped on at nearly 3,500 miles an hour. After traveling 125 miles, the spent rocket fell into the Baltic Sea.

Hitler was unimpressed with rockets so long as victory appeared certain, but when the fortunes of war turned against him, he resorted to them in desperation. A crash program was initiated and in a relatively short time, a considerable number of rockets were produced and launched successfully. The outcome of the war might have been different and some American cities might have been subjected to rocket attacks had he and his advisers recognized the military potential of rockets earlier in the war.

The Allied Forces bombed Peenemünde causing great destruction which, together with the severe shortages of materials and fuels, severely retarded the German rocket program. In spite of this, they produced about 900 V-2 missiles per month and during September 1944, launched the first V-2 bomb to fall upon England.[19] Up to 1945, England continued to be bombed by both V-1 (buzz bombs) and V-2 bombs with more than 8,000 V-1 bombs being aimed at London of which 2,420 reached it. It has been estimated that 4,300 V-2 bombs were fired during the closing stages of the war. Of these 1,500 were directed at England of which more than 1,100 reached the island. By late March 1945, the German forces had been driven back from the continental coast and England ceased to be the primary V-2 target.

The German V-2 rockets became war prizes at the conclusion of the war and the unity of the Allied Forces gave way with each country endeavoring to obtain as many of the V-2 rockets as possible.[20] The United States forces took possession of 100 of these rockets and a large quantity of parts. In addition, many of the German rocket scientists surrendered to the United States forces rather than to the Soviet forces and were later brought to the United States to participate in its rocket program. The Soviet forces captured several rocket plants and a number of rockets which, together with the help of other German rocket scientists, enabled the Soviet Union to produce a number of rockets in occupied Germany. This gave the Soviet government considerable advantage in that it had an expandable supply of V-2 rockets for experimentation, whereas the United States was limited to those it had seized.[21]

REFERENCES

1. Andrew G. Haley, *Rocketry and Space Exploration,* (New York: D. Van Nostrand Company, Inc., 1958), p. 44.
2. Beryl Williams and Samuel Epstein, *The Rocket Pioneers, On the Road to Space,* (New York: Julian Messner, Inc., 1958) pp. 52-69. See: Karl Gilzin, *Sputnik and After,* (Translated by Pauline Rose) (London: MacDonald 1959, for a general coverage of Tsiolkovskii's work relating to the development of rocketry in the Soviet Union.
3. *Ibid.*, pp. 70-110.
4. *Smithsonian Miscellaneous Collections,* Volume 71, (1919-1927) No. 2, pp. 1-69.
5. *Ibid.*, pp. 58-59.
6. *Ibid.*, p. 56.
7. *The New York Times,* January 12, 1920, pp. 1-3.
8. *The New York Times*; January 13, 1920, "Topics of the Times," p. 12.
9. The Report of the flight was not published until some years later.
10. *The New York Times,* July 18, 1929, p. 2.
11. *Smithsonian Miscellaneous Collections,* Volume 95 (1936-37), No. 3, pp. 1-10 plus eleven plates. (The report was made to the Daniel and Florence Guggenheim Foundation.)
12. Beryl Williams and Samuel Epstein, *op. cit.*, pp. 111-143. Also: Hermann Oberth, (Translated by G. P. H. de Freville) *Man Into Space,* (New York: Harper & Brothers, Publishers, 1957), pp. vii-ix.
13. Hermann Oberth (Translated by G. P. H. de Freville) *Man Into Space,* (New York: Harper & Brothers, Publishers, 1957), pp. vii-ix.
14. Released on October 15, 1929 in Berlin.
15. Willy Ley, *Rockets, Missiles and Space Travel,* (Rev. Ed.) (New York: The Viking Press, 1957) Appendix 2, pp. 443-445. (Addendum to Chapter 6)
16. Beryl Williams and Samuel Epstein, *op. cit.*, pp. 144-170. Also: Willy Ley, *Ibid.*, pp. 131-139.
17. *Ibid.*, p. 172. (Also: Willy Ley, *op. cit.*, pp. 443-444) (Correct date 1638)
18. Wernher von Braun later became one of the outstanding rockets experts in the United States.
19. Beryl Williams and Samuel Epstein, *op. cit.*, pp. 227-228.
20. Martin Caidin, *Countdown for Tomorrow,* (New York: E. P. D'utton and Company, Inc., 1958), p. 128.
21. *Ibid.*, p. 208.

Chapter II

SPECTACULAR SPACE FLIGHTS BY THE SOVIET UNION 1950's - 1960's

Mankind will not stay on Earth forever, but, in the pursuit of the world and space, will at first timidly penetrate beyond the limits of the atmosphere and then will conquer all the space around the sun.

Konstantin Eduardovich Tsiolkovskii (1913)[1]

It was from the soil of the Union of Socialist Republics (USSR) that the first artificial satellite was thrust into orbit about the earth on October 4, 1957.[2] The launching of Sputnik I into an earth orbit was one of the great scientific and technological achievements of modern times setting a new timetable for interplanetary travel and making more acceptable the possibilities of space travel.

Sputnik I was an unmanned spacecraft carrying few instruments and causing great consternation on the part of Western space scientists, especially those of the United States. Regardless of the circumstances under which it was launched and the propaganda attached to it, the Soviet Union did achieve a scientific breakthrough.

The satellite was a shiny globe having a diameter of 22.8 inches, weighing 184 pounds and equipped with four antennas extending outward from it. It was assumed that it would remain in earth orbit for several years, but in early January 1958, it reentered the earth's atmosphere and disintegrated somewhere over Mongolia. Tass reported that it had covered about forty million miles during the time it was in space.

Sputnik I was thrust into an earth orbit by a three-stage rocket. The first stage burned for one or two minutes accelerating the rocket to between 4,350 and 4,661 miles an hour. About a mile up, the rocket inclined from the vertical and, after the first stage separated from it, moved at 45 degrees to the surface of the earth. The second stage ignited, increasing the speed of the third stage to between 11,185 and 12,428 miles an hour, and with the satellite continued upward on its own momentum until it was traveling parallel to the earth's surface. About six hundred and twenty-five miles from the launching site, the third stage ignited and boosted the speed of the satellite to approximately 18,000 miles an hour, putting it into an earth orbit inclined at 65 degrees to the equatorial plane. The apogee of the orbit was 560 miles and the perigee 145 miles, requiring 1 hour and 36.2 minutes for its completion.

The world had not recovered from the shock and surprise of this breakthrough when on November 3 of the same year, the Soviets fired unmanned Sputnik II into an

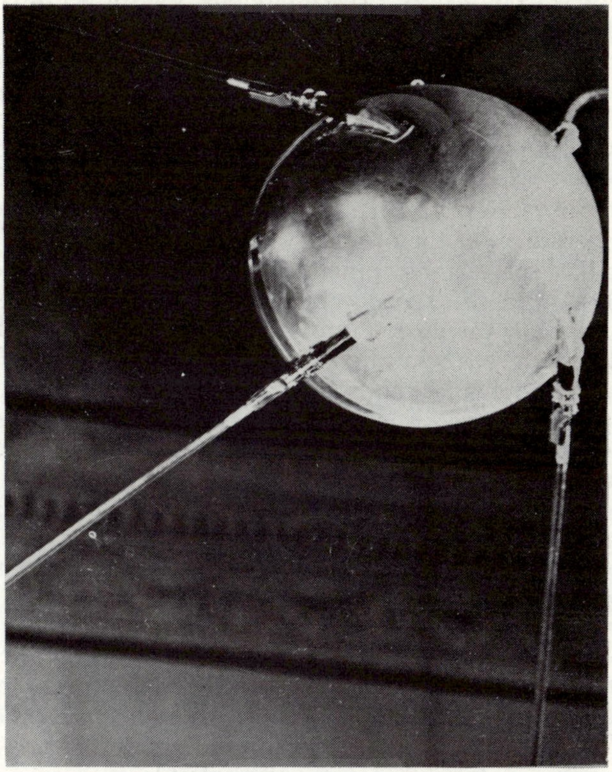

Sputnik I, the First Man-Made Earth-Orbiting Satellite, October 4, 1957 (Source: NOVOSTI from Sovfoto)

earth orbit, with an apogee of 1,056 miles and a perigee of 150 miles. The speed of the satellite was about 17,895 miles an hour allowing it to circle the earth in 103.7 minutes. Sputnik II continued in orbit until April 14, 1958 when it disintegrated upon reentering the earth's atmosphere.

The 1,120.3-pound Sputnik II was conical in shape, measuring four feet in diameter and 19 feet in length and capable of carrying a payload six times greater than the total weight of Sputnik I. Its mission, as stated, was to collect and transmit to earth data relating to cosmic rays, temperatures and pressures in the upper layers of the earth's atmosphere and in space. Inside the capsule was a small Eskimo dog (Kudryauka) hermetically sealed in an air-conditioned compartment. It had attached to it equipment for recording its reactions under prolonged conditions of weightlessness and for feeding it. No provision appears to have been made for the safe reentry of the satellite into the earth's atmosphere and its recovery. *Sputnik II was the first spaceship put into an earth orbit with an animal aboard.*

On May 15, 1958 the Soviet Union launched unmanned Sputnik III weighing 2,925.53 pounds and measuring 5 feet 8 inches in diameter and 11 feet 9 inches in length.[3] Its earth orbit had an apogee of 1,168 miles and a perigee of 150 miles, requiring 106 minutes to complete. No safeguards were made for protecting the satellite when it reentered the earth's atmosphere or for its ultimate recovery. The life of the spaceship was estimated to be six months.

The capsule was designed as a space laboratory with 2,129 pounds of scientific instruments and electronic devices operated by regular batteries, supplemented by solar batteries. It carried no living matter and consequentially, no provision was made for conducting biological experiments. There were three major groups of devices: Group I: instruments, designed to measure solar radiation, cosmic rays, micrometeorites, Group II: devices for studying the composition of the atmosphere, pressure ionization, electrical phenomena, and the magnetic field of the earth, and Group III: a powerful radio transmitter, temperature regulators and a control svstem to program the operations of the other instruments.

The spectacular space achievements of the Soviet Union during 1959 were moon-probing shots and the space vehicles used were referred to as Luniks. On January 2, 1959, Lunik I was hurled into space aimed at the moon. The last stage of the rocket and the lunar spacecraft had a combined weight of 3,238 pounds exclusive of fuel. Of this total weight, 795 pounds consisted of instruments for measuring the magnetic field of the moon, cosmic radiation, gaseous components of interplanetary dust, and the corpuscular radiation of the sun. It was not revealed whether the spacecraft was to impact the surface of the moon or go into lunar orbit. In the end, it did neither, but passed within 4,660 miles of the moon at a speed of 1.5224 miles a second and went into orbit about the sun, *becoming the first man-made space vehicle to orbit the sun.* About sixty hours after the spacecraft was launched, it had traveled 373,125 miles and its radio had stopped transmitting.

Lunik II, a spherical capsule, was launched on September 12, 1959, by a guided missile, the final stage weighing 3,330 pounds exclusive of the fuel. The spacecraft's scientific equipment and measuring devices together with the power installation, including the container, weighed 867 pounds. In addition to the scientific equipment, it carried a pennant and the hammer and sickle of the Soviet Union. The sphere was hermetically sealed to prevent the possible contamination of the surface of the moon with earth microorganisms. This precaution was taken to ensure that, when man finally landed on the moon, he would find it in its natural and uncontaminated state. The capsule impacted the moon about thirty-five hours after being launched, *marking the first time that an object had been sent from the earth to another celestial body.*

On October 4, 1959, the second anniversary of Sputnik I, Lunik III was launched into space. The capsule was an automatic interplanetary station designed to partially encircle the moon, and then go into orbit about the earth. *This marked the first time that photographs of the hidden side of the moon were transmitted to earth.*

On May 15, 1960, the Soviet Union launched a 10,005-pound spaceship with a dummy cosmonaut aboard into earth orbit with an altitude of 500 miles, requiring 91 minutes for its completion. The primary objectives of the mission were to test the spacecraft's systems for safe flight, control, reentry requirements, and the conditions essential to the comfort and survival of the crew, for the time when manned spaceships would be sent into space. The pressurized cabin weighed 2.5 tons and was designed to separate from the carrier rocket and orbit about the earth. It was equipped with a radio transmitter and special equipment for playing back data obtained by detecting instruments powered by chemical and solar batteries. Later, an attempt was made to return the capsule to earth but the retrorockets malfunctioned causing it to assume an elliptical earth orbit.

A five-ton spaceship, carrying two dogs and a television camera, was launched on August 19, 1960, and went into a near-circular orbit about the earth with an elevation of 199 miles an inclined at 65 degrees to the plane of the equator. The period required for the spaceship to complete the orbit was 90.6 minutes during which time it approached the fringes of the polar regions. Two dogs, *Stryelka* (Little Arrow) and *Byelka* (Squirrel) with white fur, making for better observations by television, were aboard the space capsule, and were confined in individual glass compartments with the instruments attached to them. In addition, the capsule carried other organisms such as rats, mice, flies, plants, seeds, and fungi.

During the period of weightlessness, the dogs reacted differently. One appeared more excited and breathed more rapidly than the other but later resumed a more normal breathing rate. Electrocardiograms and physiological examinations of the dogs, after they had returned to earth, revealed no ill effects from the space journey. The other living organisms appeared to have sustained themselves during the cosmic flight and one plant continued to blossom after its return to earth. One of the significant aspects of this particular mission was the successful recovery of the undamaged space capsule and the safe return to earth of its living cargo with no apparent ill effects. The capsule was reported to have landed on August 20, 1960, within six and a quarter miles of the planned landing area, indicating that the Soviet Union had resolved the reentry and safe-landing problems. *This was the first time that living organisms on board a spaceship had been returned to earth from space.*

On December 1, 1960, another five-ton spacecraft was placed in orbit about the earth carrying two dogs, *Pchelka* (Little Bee) and *Mushka* (Little Fly), and other living organisms. The satellite assumed an orbit with an apogee of 159 miles and a perigee of 117 miles. This orbit was too low to permit the spaceship to remain long in space because of the gravitational pull of the earth. Unfortunately, the spacecraft went out of control and disintegrated upon reentering the earth's atmosphere.

Sputnik V, weighing, exclusive of the last stage, about 7.1 tons, *the heaviest spaceship to be placed into an earth orbit up to that time,* was launched on February 4, 1961, by a multi-stage rocket. The satellite assumed an earth orbit with an apogee of 204 miles and a perigee of 140 miles and was inclined at 65 degrees to the plane of the equator. Apparently, the spaceship was not equipped for reentry nor did it have living organisms aboard.

On February 12, 1961, an interplanetary capsule Venera I, weighing 1,416 pounds was sent on a mission to Venus with the expectation that it would reach the vicinity of that planet by May. *This was the first satellite to be launched from an earth-orbiting station, parked in space,* and by noon of that day, was reported to have traveled some 78,500 miles. It finally passed within about 60,000 miles of the planet.

Another five-ton Soviet spacecraft was launched on March 9, 1961 carrying a dog named *Chernushka* (Blackie) and other living organisms, to be observed by television during the flight. It brought the dog back safely to earth without any apparent ill effects.

Subsequent to May 15, 1960, the Soviets had placed four five-ton spacecraft into earth orbit and again on March 25, 1961, launched another earth-orbital spaceship, similar to that launched on March 9, carrying another dog named *Zvezdochka* (Little Star). The apogee of its orbit was 154 miles and the perigee 109 miles, requiring 85 minutes and 42 seconds to complete. Again, the satellite and the dog returned to earth safely.

April 12, 1961 marked the launching of Vostok I (East), a manned five-ton spaceship carrying Major Yuri Alekseyevich Gagarin,[4] *the first man to travel in space.* The space capsule orbited the earth in 89.1 minutes on a path with an apogee of 188 miles and a perigee of 109 miles. During the 108 minutes in space, he maintained constant radio, telemetric, and television contacts with ground control stations. The flight was terminated after the completion of one orbit at a speed in excess of 17,000 miles an hour. Following reentry of the space capsule into the earth's atmosphere, a large parachute attached to it was deployed allowing it to float safely to the surface of the earth.

Less than four months after Major Gagarin's mission, Major Gherman Stepanovich Titov traveled into space on August 6, 1961 in the 10,427-pound spacecraft, Vostok II. It assumed a path about the earth with an apogee of 160 miles and a perigee of 111 miles, requiring 88.6 minutes to complete. Major Titov circled the earth more than 17 times in 25 hours and 18 minutes, during which time ground control stations observed him by television. He took over the spaceship's manual control system during the fourth revolution and piloted it for an hour. During the eighteenth revolution, after traveling some 435,000 miles, he reentered the earth's atmosphere and landed safely.

The third cosmonaut to orbit the earth was Andrian G. Nikolayev, who was hurled into space in Vostok III on August 11, 1962. The orbit assumed by the spaceship had an apogee of 145 miles and a perigee of 112 miles and required 88.3 minutes, traveling at about 18,000 miles an hour, to complete. On the following day, Lieutenant Colonel Pavel R. Popovich was sent into space in Vostok IV and by noon, made a rendezvous with Vostok III, but apparently the spacecraft did not make physical contact or dock.

A half an hour after the rendezvous, the spacecraft were flown in group formation only a short distance apart. The cosmonauts, as they ate, exercised, and slept, were observed on television by ground station personnel and the public. *This was the first live television broadcast from a spaceship while in earth orbit*. As of August 13, Major Nikolayev in Vostok III had completed 31 earth orbits and had traveled about 776,750 miles. The two cosmonauts landed safely on August 15, Major Nikolayev landing at 9:55 a.m., Moscow time after completing 63 orbits of the earth, and Colonel Popovich at 10:02 a.m. after completing 47 orbits. *The two cosmonauts had exceeded all previous records for extended manned space flights and had traveled more than a million miles.*

An unmanned spaceship, Mars I, was launched on a Mars-probing mission by the Soviet Union from an earth-parking orbit on November 2, 1962. It was estimated that it would take seven months for the spaceship to arrive in the vicinity of that planet. The satellite weighed about 1,970 pounds and was equipped with telemetric measuring and scientific instruments, activated while in space by radio signals from ground stations. The purpose of the mission was to collect and transmit to earth photographs and other information about Mars and provide evidence, if any, as to the existence of some form of Martian life. *This was the first launching of a Mars probe designed to transmit to earth radio photographs of that planet*.

Lunik IV, an automatic station weighing 3,135 pounds, became, on April 2, 1963, the fourth spaceship launched to gather information relating to the moon. The vehicle was placed in an earth-parking orbit from which it was ejected and sent on an estimated three and a half day-journey to the vicinity of the moon. No advance information was given, before or at the time of launching, as to whether it was to pass by the moon or crash-land.

On June 15, 1963, Lieutenant Colonel Valery F. Bykovsky, in Vostok V, became the fifth cosmonaut of the Soviet Union to be placed in an earth orbit. For about ninety minutes, the people of the Soviet Union and Eastern Europe viewed on live television, Colonel Bykovsky in the cabin of his spaceship as it circled the earth. During the first 24 hours, he traveled more than 416,000 miles and completed 16 orbits of the earth. For a brief period, he took over the manual controls and guided the spaceship on its path around the earth.

Forty-five and a half hours after the launching of Vostok V, Junior Lieutenant *Valentina V. Tereshkova, was launched into an earth orbit in Vostok VI and became the first woman to travel in space.* The earth orbit assumed by the spaceship was elliptical with an apogee of 144 miles and a perigee of 112 miles, requiring 88.3 minutes to complete. In contrast, the apogee of Vostok V's orbit was 129 miles and the perigee 105 miles, taking 88.06 minutes to complete. Lieutenant Tereshkova was observed also by live television as the two spacecraft orbited the earth in group flight about three miles apart.

The two space travelers were returned safely to earth on June 20, 1963. Lieutenant Colonel Bykovsky was uninjured but Junior Lieutenant Tereshkova suffered a slight bruise on the nose. During his flight in space, Lieutenant Colonel Bykovsky completed 81 orbits exceeding the record of 17 orbits established by Major Titov, and traveled in 4 days, 23 hours and 6 minutes, over 2,000,000 miles. Junior Lieutenant Tereshkova completed 48 orbits about the earth in 70 hours and 50 minutes.

It is suspected that the Soviet Union experienced two failures in attempting to send unmanned artificial satellites to Venus during February and March, 1964. On April 2, 1964, Zond I, a new type of unmanned automatic station, was ejected from an earth-parking orbit and sent on an interplanetary mission with an undisclosed destination. It was generally believed that the mission had as its destination Venus, and that it was unsuccessful in achieving its planned goal.

An outstanding Soviet space shot was executed on October 12, 1964, when Voskhod I (Sunrise), carrying three cosmonauts, was placed in an earth orbit with an apogee of 254 miles and a perigee of 111 miles, taking 90 minutes to complete. This apogee was 52 miles higher than that of the spaceship which carried Colonel Garagin aloft.

The purpose of the mission was to test the new multi-seat spaceship, check the capacity for work and interaction during space flight conditions of the group of cosmonauts, carry out scientific physical and technical investigations in space, continue the study of the effects of different factors on man's organism, and conduct extensive medical-biological research under conditions of weightlessness. Colonel Vladimir M. Komarov was in command and was accompanied by Konstantin P. Feoktistov, a designer-scientist, and Lieutenant Boris B. Yegorov, a physician. The new troika was attired in lightweight steel grey pants, blue jackets and white helmets with earphones. After the spaceship had completed 16 earth orbits, covering about 437,000 miles, in 24 hours and 17 minutes, the cosmonauts, *manning the world's first multi-seat spacecraft*, requested that they be allowed to remain in space for another twenty-four hours. Their request was denied and the mission terminated on October 13, with their safe return to earth.

November 30, 1964 marked the launching of an unmanned Soviet spaceship, Zond II, on a trajectory leading to Mars. It was reported that it was equipped with a new type rocket engine known as a plasma accelerator. The trip to Mars was estimated to take seven and a half months. During the initial stages of the flight, it was reported that the spacecraft experienced partial power failure but later stable radio contact was maintained. It is questionable whether the mission was successful in achieving its intended goal.

A spectacular launching was achieved on March 18, 1965, when Voshkod II was placed into earth orbit by a launch rocket believed to have the capability of developing 900,000 pounds of thrust. The apogee of the orbit was about 309 miles and the perigee about 108 miles requiring 90.9 minutes to complete. The maximum altitude of the orbit attained, in this instance, was greater than that attained by manned spaceships launched previously by the Soviet Union.

Cosmonauts Colonel Pavel I. Belyayev (Pilot), and Lieutenant Colonel Aleksei A. Leonov, who were aboard Voshkod II, became the tenth and eleventh Soviet spacemen to orbit the earth. They were brought safely back to earth after completing 17 orbits and being in space for 26 hours. *During the flight, Lieutenant Colonel Leonov left the spaceship and floated in space while attached to it by an umbilical cord. He remained outside for ten minutes and was the first human being to float in space.* Colonel Leonov was observed on live television by millions throughout the Soviet Union and Eastern Europe. They saw him leave the spaceship's cabin, raise his hands, pull up his legs and extend them outward while holding onto a handrail, push himself away from the spaceship and perform other movements. He had no device by which to direct his movements while in space should the umbilical cord break. If this had happened, he would have drifted in space and ultimately would have reentered the earth's atmosphere at such a high rate of speed that his body would have disintegrated because of the intense frictional heat.

During the space walk, Colonel Leonov wore a specially designed multi-layer pressurized spacesuit, gloves and footwear, described as a miniature hermetic cabin. In addition to the specifically designed suit equipped with its own power supply, communication system, disposal system and air conditioning, he wore a metal helmet with a transparent visor. If the sun's rays had penetrated any part of the suit, he would have died instantly.

Luna V, an instrumented space station weighing 3,254 pounds, launched on May 9, 1965, was unique in that it was the first announced Soviet moon mission in a two-year period. Unfortunately it crash-landed on May 12 and became the seventh man-made earth object to impact the surface of the moon.

Luna VI, launched on June 8, 1965, was apparently another attempt to soft-land an instrument-carrying capsule on the surface of the moon. However, on June 10 it was announced that it would miss the moon by 100,000 miles because of the failure of an onboard motor, used to adjust the trajectory in midcourse.

The unmanned spacecraft, Zond III, was launched on a solar-orbital mission on July 18, 1965 to test its automatic rocket system under conditions of prolonged flight and to carry out interplanetary scientific studies. On July 19, it was reported to have traveled 140,000 miles from earth and to be transmitting back to earth scientific data. It transmitted photographs of the far side of the moon, some of which were taken less than 7,000 miles above its surface.

On October 4, 1965, efforts were renewed to soft-land an instrumented package on the lunar surface, when Luna VII, weighing 3,302 pounds, was projected into space by a powerful multi-stage rocket. The following day, it was reported that the moonshot had failed in soft-landing the capsule on the lunar surface, and had crashed on the Ocean of Storms located in the western half of the moon's visible hemisphere.

A fourth attempt to soft-land an instrumented capsule on the surface of the moon was made when Luna VIII was launched on December 3, 1965. Once again, something went wrong, during the final stage of touchdown, causing it to crash-land, this time on the Sea of Storms in the eastern portion of the visible side of the moon.

The Soviets launched two spacecraft during 1965 to explore Venus. On November 12, 1965, the 2,122-pound Venera II was fired into space followed by Venera III on November 16. The planet was, at the time of launching, 38 million miles from the earth and it was estimated that it would take about three and a half months for the spacecraft to traverse the distance. Venera II passed at a distance of about 25,000 miles of Venus on February 27, 1966, but failed in its mission to photograph the planet. Venera III crash-landed on Venus on March 1, *which was the first physical contact with that planet by man.* However, the spacecraft failed to eject its 35-inch sphere designed to measure the temperatures of the planet.

Luna IX, one of the five moon-probing satellites, sent into space by the Soviet Union during 1966, was launched on January 31, of that year and soft-landed on the Ocean of Storms on February 3. *It became the first instrumented capsule to transmit close-up pictures of the lunar surface and the first space object to soft-land intact on the moon.*

Unmanned Luna X, a 540-pound spaceship, was fired into space on March 31, 1966, and went into a lunar orbit with an apolune of 621 and a perilune of 217 miles, requiring three hours to complete. Its mission was to test the systems of the lunar-orbiting satellite for near-lunar observations. As of May 3, it had completed its first month in a lunar orbit, made 244 lunar revolutions, and continued to circle the moon with all systems operating. The capsule stopped transmitting on May 30, after having circled the moon 460 times. Up to that time, 219 sessions had taken place between it and ground stations.

On August 24, 1966, unmanned Luna XI was launched and on August 28 went into lunar orbit, with a apolune of 745 miles and perilune of 99 miles, requiring 2 hours and 58 minutes to complete. The satellite weighed 3,608 pounds and was more fully instrumented than previous Soviet lunar spacecraft. As of September 14, it was still orbiting the moon and transmitting but on October 4, it was announced that its mission had been terminated and the final radio signals had been received from it on October 1.

Luna XII was launched on October 22, 1966, and assumed a lunar orbit with an apolune of 1,078.8 miles and a perilune of 62 miles, taking 3 hours and 25 minutes to complete. Its lunar mission was to test the systems of the spacecraft and conduct further exploration of the moon and its environment. It sent back pictures of the flat portion of the Sea of Rains taken from an altitude of 62 miles above the surface covering an area 19.22 square miles, of which two were shown on the Moscow television. By November 26, it had circled the moon 200 times, traveling 1.8 million miles in space. Important information was also transmitted relating to X-ray and gamma radiation on the moon's surface, the intensity of the sun's radiation, and micrometeorite activity in the near-lunar space.

Another lunar-probing satellite, Luna XIII, was successfully launched on December 21, 1966, and on December 24, soft-landed on the moon on the Ocean of Storms. The spaceship was similar to that used for the Luna IX mission except it had two slender booms, extending about five feet, for measuring density and firmness of the moon's surface. Four minutes after landing on the Ocean of Storms, it began transmitting signals. The photographs sent back revealed the moon's surface as being stony and uneven and subject to shifts and erosion. The tests of the lunar surface showed it to be firm and its soil to be of medium density to a depth of eight to 12 inches. This lunar mission was terminated on December 30, 1966.

Soyuz I (Union), carrying Colonel Vladimir K. Komarov, on his second journey into space, was hurled into earth orbit on April 22, 1967, with an apogee of 140 miles and a perigee of 125 miles requiring 88.6 minutes to complete. Later it was placed in a lower orbit with an apogee of 137 miles and a perigee of 124 miles. The stated purposes of the mission were to test the spacecraft, carry out scientific and technical experiments, and conduct certain biological studies. Shortly after becoming spaceborne, Colonel Komarov radioed a message commemorating the 50th Anniversary of the Great October Socialist Revolution. By the end of 23 hours and 36 minutes in space, he had completed 16 orbits of the earth during which time he slept for about eight hours while the spacecraft was out of radio range with ground stations. Colonel Komarov was fatally injured on April 24, when, after reentry into the earth's atmosphere, the capsule's main parachute snarled causing it to plunge 4.3 miles to earth. *This was the first known fatality in connection with the Soviet Union's manned space flights.*

Venera IV was launched on June 12, 1967, on a four-month journey to Venus. The 2,437-pound spacecraft, described as an automatic research station carried a roly-poly type capsule, coated with a coked layer, to control its internal temperature through evaporation when subjected to the intense heat of Venus. After reaching the vicinity of the planet on October 18, the capsule weighing 2.5 pounds and equipped with sensoring devices, but no camera, was ejected and descended through the Venusian atmosphere to impact the surface of the planet. As it dropped 15.5 miles to the planet's surface, it transmitted data for about one and a half hours, confirming that the atmosphere of Venus consisted principally of carbon dioxide, except for about 1.5 per cent hydrogen and traces of water vapor. There was no noticeable trace of nitrogen. The atmosphere was found to be 15 times denser than that of the earth with temperatures ranging between 40 and 280 degrees Celsius (104 to 536 degrees Fahrenheit). The planet appeared to have a weak magnetic field, no radiation belt, and a weak hydrogen corona. From the data, it may be concluded that there are no living organisms on Venus such as exist on the earth. *This was the first time that a working instrumented capsule had been landed on another planet.*

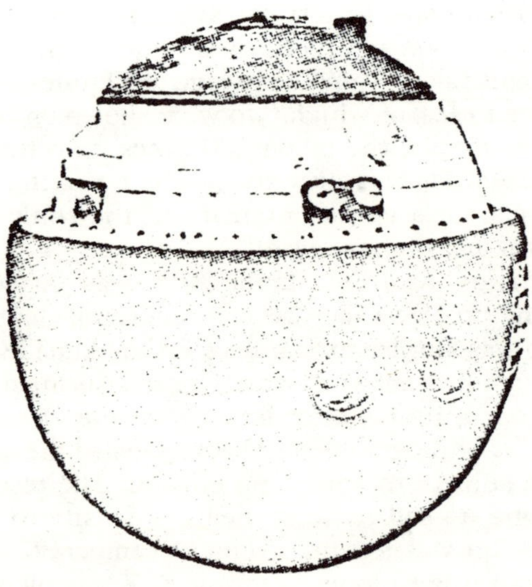

Soviet Capsule which Landed on Venus after being Ejected by Venus IV Spaceship (Source: New York Times Index, October 1967)

Two unmanned cosmos spacecraft were launched into separate earth orbits on October 30, 1967 and docked automatically by command signals from the earth.[5] Cosmos 186 was first placed in earth orbit on October 27 and Cosmos 188 was launched on October 30 at the precise time to enable it during its first orbit to dock with Cosmos 186 when it was in its forty-ninth orbit. When Cosmos 188 entered its orbit, the two spacecraft were only 14.9 miles apart. After docking, they flew as a unit for three and a half hours in an orbit with an apogee of 171 miles and a perigee of 124 miles, requiring 88 to 97 minutes to complete. After undocking, Cosmos 186 returned to earth on October 31 while Cosmos 188 continued in earth orbit. *This was the first automatic docking and undocking of unmanned spacecraft while in earth orbit by the Soviet Union.*

Zond IV, an automatic research station, was injected into interplanetary space from an earth-parking orbit on March 2, 1968. Its stated mission was to study the outlying regions of near-earth space. Little other information was given as to its objectives or destination.

On April 7, 1968, Luna XIV, described as an unmanned automatic space station, was sent into space and assumed a lunar orbit with an apolune of 540 miles and a perilune of 100 miles, requiring 2 hours and 40 minutes to complete. Its primary purpose was to carry out scientific studies in near-lunar space.

Unmanned Cosmos 212 and Cosmos 213 spacecraft were placed in earth orbits to test docking maneuvers under automatic control on April 14 and 15, 1968, respectively. The spacecraft located each other by radar and succeeded in docking while traveling above the Pacific Ocean. They flew together for 3 hours and 50 minutes, before returning to earth safely. This was the Soviet Union's second automatic docking in less than six months. Cosmos 212 landed on April 19 and Cosmos 213 on the following day.

On September 15, 1968, Zond V, a new unmanned outer space-probing vehicle, was launched on an interplanetary mission to test its systems and carry out certain biological and botanical experiments. On September 17, Jodrell Bank Observatory in England reported that it had passed within 1,000 miles of the moon and was believed to be on its way back to earth. A Soviet announcement came two days later stating that it had been launched on September 15, flown within 1,200 miles of the moon, had completed its mission and was continuing to relay scientific information. On September 23, it was announced that Zond V had been successfully guided back to earth on September 22, and had been recovered from the Indian Ocean. *This was the first recovery outside the borders of the Soviet Union, and was the first spacecraft to orbit the moon and return to earth safely.*

Soyuz II and III were put into space during the autumn of 1968 to test rendezvous procedures between an unmanned and a manned spacecraft. Soyuz II, the unmanned spacecraft, was launched on October 25, 1968 followed on October 26 by the Soyuz III, piloted by Colonel Georgi T. Beregovoi. The initial stages of the rendezvous maneuver were carried out by an automatic system, after which Colonel Beregovoi took over and completed the operation manually. Later, a second rendezvous was achieved between the two spacecraft, but it was not revealed whether they had actually docked. Soyuz II was brought safely back to earth in a soft-landing on October 28 and Soyuz III on October 30 after nearly four days in space.

On November 10, 1968, Zond VI, an unmanned spacecraft was placed in an earth-parking orbit from which it was sent into interplanetary space. This was the second moonshot in less than two months and was reported to be carrying live organisms. Jodrell Bank Observatory in England, reporting on its flight stated that it had approached to within 1,000 to 2,000 miles of the moon, circled it, and was heading to earth. This was confirmed by a Soviet news release on November 14 stating that the spacecraft had come within 1,500 miles of the moon, after which it was guided to a soft-landing on Soviet territory on November 17.

Venera V and Venera VI, two 2,500-pound unmanned space capsules, were launched from an earth-parking orbit toward Venus on January 5 and 10, 1969, respectively. It was reported on May 16 that the former had completed its four-month mission and had transmitted for 53 minutes scientific information about the Venusian atmosphere with its thick clouds, torrid heat, and abundant carbon dioxide. It was revealed that the capsule descended through the planet's atmosphere attached to a parachute and that it contained a portrait of Lenin and the Soviet coat of arms. Venera VI entered the Venusian atmosphere on the following day and transmitted information for 51 minutes while floating downward. No information was given as to the landing of these capsules and it is not known whether they soft-landed or crashed upon the surface of Venus.

The Soviet Union achieved a successful docking of two manned spacecraft, Soyuz IV and Soyuz V. Soyuz IV was launched on January 14, 1969, carrying Lieutenant Colonel Vladimir A. Shatalov and Soyuz V on January 15, carrying Colonel Boris Volynov commander, Aleksei Yeliseyev flight engineer, and Lieutenant Colonel Yevgeny Khrunov research engineer. On January 16, Cosmonaut Shatalov alone in Soyuz IV searched out and linked up with Soyuz V using the manual controls, and the two spacecraft remained joined for 4 hours and 35 minutes. Aleksei Yeliseyev and Lieutenant Colonel Yevgeny Khrunov space-walked from Soyuz V to Soyuz IV and remained aboard with Colonel Vladimir Shatalov. Soyuz IV returned safely to earth on January 17, and Soyuz V on the following day. *This was the first manned docking of two manned spaceships and transfer of personnel from one spaceship to another in space.*

The Soviet Union's unmanned lunar spaceship Luna XV, launched on July 13, 1969, three days before the scheduled lift-off of Apollo 11 spacecraft caused considerable uneasiness on the part of some space scientists of the United States. Information relating to the mission was not released until some time after lift-off and then only that it was to study the moon and near-moon space. The post-lift-off information came primarily from the Jodrell Bank Observatory in England. As a result of this secrecy, there occurred considerable speculation as to the mission's purpose. It was thought by some that it might be to scoop automatically a sample of lunar soil and bring it back to earth for analysis.

Colonel Frank Borman,[6] who had previously visited the Soviet Union, requested assurance from the President of the Soviet Academy of Science on July 17, 1969, that the mission would not interfere with the Apollo 11 mission. This assurance was received in a telegram from the President of the Academy of Science who informed him that Luna XV was in moon orbit with an apolune of 127 miles and a perilune of 35 miles,[7] and would advise him if Luna XV's flight path were changed. Sunday night it was announced that its orbit had been changed bringing it within ten miles of the moon.

The Jodrell Bank Observatory reported on July 21, that Luna XV had landed on the Sea of Crises about five hundred miles from where Apollo 11 had landed the day before. No statement about the landing was made by the Soviet authorities, and it was assumed by Western space scientists that it had crashed on the surface of the moon.

The unmanned Zond VII was fired moonward on August 8, 1969 with the usual concise statement that its mission was to study the moon and near-lunar space. Much was made of the fact that it was propelled by a powerful carrier rocket causing some speculation that there might be an attempt to land the spacecraft on the moon. Apparently it was not equipped to do so and instead circled the moon using the lunar gravity to draw it around the moon, after which it assumed an earth trajectory. It was reported that, while circling the moon, photographs were taken of the moon and the earth and measurements were made of the physical characteristics of the near-lunar space and other scientific phenomena. The mission was terminated on August 14, when the space capsule soft-landed in Northern Kazakhstan in the Soviet Union.

Considerable interest and speculation by Western space scientists centered on the successive launchings of three manned Soyuz spacecraft during October 1969, demonstrating the Soviet Union's capability for placing spacecraft into earth orbits. On October 11, Soyuz VI, carrying Lieutenant Colonel Georgi S. Shonin and Valery N. Kubasov, was hurled into an earth orbit. On October 12, Soyuz VII was sent into space, commanded by Lieutenant Colonel Anatoly V. Filipchenko, accompanied by Vladislov N. Volkov, a civilian back-up pilot, and Lieutenant Colonel Viktor V. Gorbatko, a research engineer. The next day, Soyuz VIII was launched and entered an earth orbit under the command of Colonel Vladimir A. Shatalov with Dr. Aleksei S. Yeliseyev flight engineer.[8] *Colonel Shatalov was placed in command of the group flight involving seven cosmonauts, the largest number in space at one time. The Soviet Union had again accomplished another first in space.*[9]

Each Soyuz spaceship weighed about 14,000 pounds and was a self-contained unit. It consisted of three compartments: two crew cabins and a third compartment in the rear in which were housed the instruments and rocket engines. The forward compartment, the orbital compartment, was spherical with a cylindrical collar at one end for docking with other spacecraft. It was in this compartment that the crew were to conduct experiments, eat, sleep, and exercise. They were to remain in the more elliptical middle compartment, protected by a heat shield, during lift-off, orbital maneuvers, and reentry into the earth's atmosphere. Before reentry, the rear and forward compartments were jettisoned, leaving the center compartment, occupied by the cosmonauts, to enter the earth's atmosphere and descend by means of a parachute to a soft-landing.

Little information was released by the Soviet authorities relating to this Soyuz group flight except that it was designed to carry out extensive scientific experiments. No indication was given as to whether the spacecraft would rendezvous and/or dock, leading to speculation that the joint mission was to establish an earth-orbital space station. It was also conjectured that the joint mission was primarily to regain lost prestige, resulting from the landing of men on the moon by the United States in July. Disappointing as it may have been to many, no spectacular space feat resulted from the joint mission.

It became evident that the possibility of establishing a long-term space station was doubtful because the orbital paths of the spacecraft were too low, and unless adjusted, they could not remain in space for more than two weeks. The Soyuz spacecraft's capability for life support was limited to about ten days, if it were occupied by three men, and to 20 days if it were occupied by two men. Furthermore, one of the spacecraft was reported to be without docking equipment. In the light of these facts, some Western space scientists concluded that the major purpose of the joint mission was to test techniques and procedures for establishing earth-orbital space stations. It was reported that the cosmonauts in Soyuz VI had conducted experiments in welding metals under conditions of weightlessness and near-vacuum, and that Soyuz VII and Soyuz VIII had been maneuvered to within a short distance of each other for the purpose of conducting observations and taking photographs to ascertain the visibility of objects at varying distances in space.

On October 15, it was reported that Soyuz VI had completed 66 orbits of the earth; Soyuz VII, 50; Soyuz VIII, 34, and that their average earth orbit had an apogee of 139.7 miles and a perigee of 124.2 miles. No intimation was given as to when the group flight would end. On October 16, Soyuz VI, without advance notice, returned to earth, while Soyuz VII and Soyuz VIII continued to orbit the earth. On October 17, Soyuz VII terminated its mission, removing the possibility of any linking of the spacecraft, and on October 18, Soyuz VIII soft-landed on the snow covered steppes of Kazakhstan bringing to an end the cosmic troika.

What had been accomplished by this joint mission was not clear to Western observers. The Soviet Union announced that the objective of the joint mission was to create the best possible opportunity for bringing nearer the time when permanent orbital stations would become operational, and that it had been accomplished successfully.

REFERENCES

1. Cited in Martin Caidin, *Countdown for Tomorrow*, (New York: E. P. Dutton and Company, 1958), p. 196.
2. Information concerning the space flights of the Soviet Union was obtained primarily from reports, based on releases by Tass, the Soviet News Agency, carried in American Newspapers and other publications, including *The New York Times* and *Aviation Week and Space Technology*. The approximate dates of news releases can be determined by the announced dates of the space flights set out in the context.

   Many of the releases by Tass, in connection with space flights, are general, evasive and superficial and provide little information of scientific value. In many instances, little if any advanced notices of flights are given and it is often difficult to determine if they were successful, not so successful or failures in relation to their planned objectives.

   Regardless of the nature and design of these news releases, many of the flights were monitored by our electronic network, Jodrell Bank Observatory in England, and other systems, and verifications of them were possible. There is no doubt that the Soviet Union has made significant advancements in the science and technology of space. Their space scientists have demonstrated ability and ingenuity and are to be commended for their outstanding accomplishments in space exploration and space travel.
3. According to the estimates of some Western space experts, the multi-stage launch vehicle had a thrust of 500,000 pounds.
4. Major Yuri A. Gagarin was killed in a training crash on March 27, 1968.
5. The Soviet Union has sent numerous cosmos spacecraft into orbit around the earth on various missions.

6. *Gemini VII Flight,* December 4, 1965, (p. 84) and *Apollo 8 Flight,* December 21, 1968, (p. 96).
7. The earth orbit was reduced to one with an apogee of 137.24 miles and a perigee of 58.9 miles on July 19. It was reduced again on July 20, to bring the Luna XV spaceship within ten miles of the moon.
8. Both cosmonauts were members of the crews of the Soyuz IV and Soyuz V joint mission in January, 1969, which accomplished the first manned docking and transfer of personnel from one spacecraft to another in space.
9. The previous record was established in January, 1969, when Soyuz IV and Soyuz V simultaneously orbited the earth carrying four cosmonauts.

Chapter III
SOVIET SPACE MISSIONS OF THE EARLY 1970's*

The early 1970's was a period of outstanding accomplishments for the Soviet Union with respect to manned earth-orbital, unmanned lunar and other planetary missions. It was also a most tragic period because of the loss of three cosmonauts. During the year 1970, it was estimated that the Russians had launched successfully seventy-four spacecraft, a third of which was launched during the last two months of the year.[1]

The solitary two-man mission of Soyuz IX, blasted off on the night of June 1, 1970, proved to be the longest manned mission up to that time. The announcement of the flight was made about a half an hour after lift-off by interrupting a scheduled Soviet radio and television broadcast. In command was Colonel Andrian G. Nikolayev, one of the original cosmonauts and the third Russian to fly in space,[2] and accompanying him was Vitaly I. Sevastyanov, a civilian aircraft engineer. The earth orbit of the spacecraft was switched after 14 orbits and 20 hours in space to an orbit with an apogee of 166 miles and a perigee of 132 miles, requiring 89.05 minutes to complete.

As usual little information was released as to the objectives of the mission, except that biological, geological and meteorological experiments would be conducted as well as testing the spacecraft's navigational control systems. Emphasis was placed on medical and biological tests of the cosmonauts to determine their abilities to withstand the effects of long periods of weightlessness. They underwent medical checks early in the flight and relayed the data to earth. During the sixteenth day in space, eye experiments were conducted to determine the effects of prolonged time in space upon their vision. On June 11, they played with the ground controllers, *the first chess match between cosmonauts in space and persons on earth,* which ended in a draw on the thirty-sixth move.

The Soyuz IX cosmonauts set a record for space endurance on June 15, having been in space for 14 days.[3] *They also established another record, when they completed more than 220orbits of the earth.* No indication was given as to when the flight would be terminated at that time but on June 19, the flight ended, when Soyuz IX landed safely about forty-seven miles west of Karaganda in Soviet Kazakhstan. The cosmonauts had orbited the earth for 17 days, 16 hours, and 59 minutes, during which time they had completed 287 earth orbits.

Venera VII an unmanned spaceship, was sent toward Venus on August 17, 1970. The capsule was designed to resist the intense temperatures and pressures to which it would be subjected when it parachuted to the surface of the planet. On January 26, 1971, it was announced officially that on December 15, 1970, after a journey of 120 days, the capsule had soft-landed on Venus. It had transmitted radio signals for about thirty-five minutes while floating down to the surface through the Venusian atmosphere, consisting of more than ninety per cent carbon dioxide. After landing, it had continued to transmit for about twenty-three minutes. *This marked the first time that scientific information had been relayed directly from the surface of another planet in the solar system to earth.* The instruments recorded and relayed to earth temperatures ranging from 847 to 923 degrees above zero (Fahrenheit), and an atmospheric pressure 90 times that of the earth. Under these conditions, it would be highly impossible for life to exist as we know it on earth. Furthermore, it is also highly improbable that human beings, clothed in advanced life support systems, could survive for long.

The exploration of the moon by unmanned spacecraft continued to be the major concern of the Soviet Union in the remaining months of 1970. During that time one Zond and two Luna missions were undertaken in September, October, and November.

Luna XVI was launched on September 12, 1970 and on September 17 entered a circular orbit of the moon. On September 19, it assumed an elliptical orbit with an apogee of 60 miles and a perigee of nine miles. It soft-landed on the Sea of Fertility in the eastern and unexplored region of the moon on September 20. The landing was monitored by the Institute for Space Research in Bocham, West Germany and by Jodrell Bank Observatory in England. Both monitors reported the transmission of fine quality television pictures.

An outstanding achievement of this mission was the automatic scooping of a sample of the lunar soil with a remotely controlled electrical drill attached to a robot arm moving horizontally and vertically on signals from the earth station. In this manner, a sample of lunar soil weighing 3½ ounces was obtained and sealed hermetically in a tube on board the spacecraft. The second notable achievement was the blasting of the spacecraft off the surface of the moon by remote control on September 21, after a stay of 26 hours. *Luna XVI was the first unmanned spaceship to make a round trip to the surface of the moon and to bring back a sample of its soil and rock, collected automatically.* The mission was terminated when the spacecraft landed successfully on September 24, in the Soviet Kazakhstan about fifty miles southeast of Dzhezhazgan. The event was cloaked with the usual secrecy but its time and place were for *the first time announced in advance.*

This mission gave support to those who argued in favor of using unmanned automatic spacecraft for exploring the moon and nearer planets of the solar system, where manned spacecraft may go at considerably greater costs, and for the exploration of distant planets where manned spacecraft may never be able to go. The mission also renewed confidence in the Soviet Union's space efforts, especially in its potential capabilities for exploration of distant celestial bodies.

Unmanned Zond VIII lifted off on October 20, 1970, for the vicinity of the moon and swung about it by a sling-shot technique rather than orbiting it. The mission objectives were to conduct scientific experiments and make measurements along the flight path and in the near-lunar space, take photographs of the moon and the earth at various distances, and check the onboard systems and the structural aspects of the spacecraft. The mission was terminated on October 27, with the splashdown of the spacecraft in the Indian Ocean about 450 miles southeast of the Changis Archipelago, or approximately 1,200 miles northeast of Mauritius Island. The capsule was recovered by a Russian ship. Landing upon water and outside of the territory of the Soviet Union was unusual, but it was explained that this was done to practice possible variants associated with returning spacecraft to earth from the direction of the Northern Hemisphere. It was stated that the spacecraft had succeeded in swinging about the moon at a minimum altitude of 690 miles, and that the objectives of the mission had been accomplished.

The unmanned Luna XVII mission was initiated on November 10, 1970 and proved more spectacular than the immediately preceding mission. Little information was given in advance as to its purpose and again there was considerable speculation by the uninformed as to its ultimate goal. The space vehicle, after orbiting the moon for two days, soft-landed on the Sea of Rains on November 17, and discharged automatically a 1,663-pound solar-battery powered lunar rover with eight wheels and remotely controlled. It descended from the spacecraft by a ramp and on reaching the lunar surface, began a series of scientific and technological investigations, before undertaking an extended excursion over the Sea of Rains. Very little information was released about its design or operation, but in appearance it looked like a pot-bellied tub with four spoked wheels on each side. It was reported to be equipped with a French laser reflector with 14-corner reflectors for measuring the continental drifts. *This was the first time that such a rover (Lunokhod) had ever been landed on another celestial body.*

Lunokhod-1. Automatic Lunar Rover Landed on the Moon by Luna 17, November 17, 1970 (Source: Wide World Photos)

The initial excursion of Lunokhod-1 was completed on November 19. It had traveled some three hundred feet over the Sea of Rains, taken photographs of the moonscape, and conducted scientific experiments, all of which were controlled remotely by radio signals from the earth. On November 21, it was parked by remote control and remained dormant for the 14-earth-day lunar night. To accomplish the parking maneuver, it took one hour and 55 minutes of remote radio communication originating on earth. There was some concern whether Lunokhod-1 would survive the intense cold of its first lunar night beginning on November 24.

Lunokhod-1 was reactivated on December 10, after having passed the lunar night, and on December 11, was moved 801 feet in and about the boulders and craters by remote control requiring nine hours of radio communication with its ground station. On December 12, it set out again and traveled 407 feet, requiring radio communication lasting for 4 hours and 27 minutes.

On December 20, Lunokhod-1 was still roaming the Sea of Rains on its tenth reconnaissance excursion, sending back pictures showing angular stones, ranging in size from 20 to 30 centimeters in diameter (8 to 12 inches) and data relating to the moon's soil. During this time, it had traveled less than a mile from its mothership before returning to it. The rover had demonstrated that it could be maneuvered over slopes as steep as 23 degrees, moved into and out of a crater 100 yards wide, and commanded to collect samples showing the physical and mechanical properties of the lunar soil. It was parked again for the second lunar night lasting from December 23, 1970 to January 7, 1971, after which it resumed its travels in exploring the Sea of Rains. At the time it was shut down, it was 1,618 yards from the original landing site.

It was announced, on October 9, 1971, that Lunokhod-1 was inoperative after ten and a half months of automatic surveillance of the surface of the moon and had stopped functioning on October 4, shortly after emerging from its eleventh lunar night. During the period of its operation, it had traveled six and a half miles over the Sea of Rains, covered an area of 800,000 square feet and transmitted to earth 200 panoramic and 20,000 single photographs. In addition, it had provided valuable information about the lunar soil, giving significant clues as to its physical and mechanical properties at 500 reference points and as to its chemical composition at 25 reference points.

The final parking of Lunokhod-1 was accomplished on September 15, 1971. It was placed in a horizontal position in order to allow the French laser reflector to continue functioning after its systems had become inoperative. In this manner the rover was converted into a fixed base station with prolonged life of about four months, when it became evident that its power was deteriorating.

The year 1971 was one in which the exploration of the planet Mars received considerable attention by the Soviet Union, because of the favorable position of that planet relative to the earth for launching Martian-bound spacecraft. This favorable opportunity would not be available again until 1986.

On May 19, 1971, the Soviet Union made its third attempt[4] to reach Mars, when it sent the 10,249 pound Mars II on a six-month journey[5] with expectations that it would reach the vicinity of that planet by November. This mission was followed on May 28, by the launching of Mars III with the same destination and with the expectation that it would reach Mars early in December. Again, because of the usual secrecy surrounding such launchings, there arose considerable speculation on the part of Western scientists as to the primary objectives of this dual mission. Mars II entered its Martian orbit on November 27 and Mars III on December 2.

It was announced on December 7, that Mars III had released a descent capsule, equipped with a television camera and other instruments and on December 2, it had soft-landed on Mars. This marked the second time that the Soviet Union had succeeded in landing an unmanned capsule on the surface of another planet which transmitted signals from the surface of the planet.[6] Video images were received from the descent capsule for 20 seconds before contact was lost. It was claimed by Soviet space officials that radio signals were received from it for three days. Mars III remained in a Martian orbit after the release of its descent capsule.

On August 24, 1972, it was reported that the missions of Mars II and Mars III had been terminated after having produced considerable information about the Martian environment, especially relating to its surface and near-surface atmospheric conditions. Its atmosphere was found to be 2,000 times drier than that of the earth and its temperatures ranged from 55 degrees above zero to 148 degrees below zero (Fahrenheit). Near its surface, the atmosphere consisted mainly of carbon dioxide, which under the impact of the ultraviolet-solar radiation, separated into carbon oxide molecules and oxygen atoms at an altitude of 60 miles. It consisted basically of atomic hydrogen at altitudes ranging from 189 to 250 miles.

Unmanned Luna XVIII, lifting off its pad on September 2, 1971, was sent on a mission with the stated objective to carry out further scientific research of the moon and near-lunar space. There was considerable speculation as to its capability because of what had been accomplished by Luna XVI and Luna XVII. It entered a lunar orbit on September 6, as reported by the Bochum Space Tracking Station in West Germany, but on September 11, Tass announced that it had failed to soft-land in the mountainous lunar region near the preplanned site on the Sea of Fertility. It was reported that the ground station had lost contact with the spacecraft at 10:48 a.m., Moscow time, believed to have been the time when it crashed on the surface of the moon. Prior to crashing, Luna XVIII had completed 54 orbits of the moon.

Luna XIX, the Soviet's last moon mission of 1971, lifted off its pad on September 28. After lift-off, it was announced that it would not land on the moon, but would continue in a lunar orbit with an altitude of 85 miles, requiring 2 hours, 1 minute and 45 seconds to complete. No announcement was made as to the purpose of the mission.

February 14, 1972, marked the initiation of the unmanned mission of Luna XX, having as its apparent purpose to land an automatic space vehicle on the moon. It was announced on February 21, that the spacecraft had soft-landed on the western side of a small crater, Apollonius C, a lunar formation west of the Greater Apollonius Crater, situated in the mountainous terrain between the Sea of Fertility and the Sea of Crises. The stated purpose of the mission was to resolve the scientific and technical problems in placing instrumented capsules safely on the rugged lunar terrain. This goal was apparently achieved, as the spacecraft landed in the area where Luna XVIII had crashed in September 1971. The site had been selected with the expectation that the rock would consist of materials ejected by Apollonius C, some 3,000 feet deep and about six miles in diameter.[7] The spacecraft returned safely to earth at night on February 26, after being on the surface of the moon for 27 hours and 40 minutes. It was recovered during a blizzard with a 1,000-foot ceiling about seventy-five miles northeast of Dzhezhagan in the Soviet Republic of Kazakhstan.

The lunar sample was removed from its sealed tube on February 27 and was found to consist of ash-colored dust with large particles. The soil was lighter in color than that returned by Luna XVI taken from the lower region of the Sea of Fertility. It contained small rounded stones resembling anorthosite at a depth of 6 to 29 centimeters (2½ to 12 inches).

Attention shifted again to Venus with the launching of the unmanned 2,600-pound Venera VIII on March 27, 1972. The space capsule was scheduled to descend through the dense and hot atmosphere surrounding that planet and to land on its surface sometime in July. The capsule had been designed to withstand the intense heat and pressure to which it would be subjected on the surface of the planet. It was announced on July 22, 1972, that Venera VIII had successfully soft-landed on the surface of Venus, after traveling for 200 million miles in 117 days and had transmitted information for 50 minutes after landing. *This was the first space capsule to land on the day side of Venus.*

The Soviet Union launched Luna XXI, an unmanned spacecraft, on January 8, 1973, with the apparent goal of placing another automatic lunar rover on the moon. As usual little information was released about the mission and on January 12, it was reported that the spacecraft had entered a lunar orbit with an apolune of 68 miles and a perilune of 58 miles at an angle of 60 degrees to the equator of the moon. The orbit was made more elliptical, bringing the spaceship within ten miles of the lunar surface. The spaceship made 41 orbits of the moon before firing its braking-rocket in preparation for a moon landing.

Doubts as to the purpose of the mission were set at rest when on January 16, 1973, Luna XXI landed on the surface of the moon and discharged Lunokhod-2 at a site on the eastern rim of the Sea of Serenity in the northeast quadrant inside the Le Monnier Crater. The landing site was 140 miles due north of the Taurus-Littrow region where Apollo 17 had landed on December 11, 1972. The region was selected because it was situated between the arid sea and the rugged uplands containing two major types of lunar terrain.

The 1,848-pound Lunokhod-2 was 185 pounds heavier and more sophisticated than its predecessor and was equipped with television cameras, a magnetometer, a drill, and a device for making chemical analyses of the lunar soil. The lunar rover was designed to operate during the extended lunar day obtaining its power from solar radiation. It remained inoperative during the long and cold lunar night. The movements of the rover were controlled by radio command signals from its earth station.

Lunokhod-2, about three hours after landing made a test run of 100 feet, taking about half an hour. It transmitted pictures of the landing module and the surrounding moonscape and within a short time, was parked to allow its solar batteries to recharge before setting out on an extended lunar excursion.

On January 17, when it was reactivated a near-accident is believed to have occurred. The initial movement of the rover was toward the landing module of Luna XXI and it threatened to impact it. Fortunately, the ground controller stopped it within 12 feet of the module's landing platform. During this incident, one of the rover's cameras took close-up pictures of the module. Apparently, the solar batteries were not recharged satisfactorily and it was reparked in a better position.

It made a test run along the Sea of Serenity on January 19 and gathered information indicating that the landing site was located in a relatively young section of the moon. The soil appeared to be at least a half a million years younger than that of the lava plains of the dry Sea of Rains explored by Lunokhod-1, landed by Luna

XVII in 1970.[8] Following several days of test runs, the lunar rover set off on its first major excursion, traveling southeast toward the Taurus Highlands about three miles from the landing site.

It was reported on February 9 that Lunokhod-2 had ended a two-week period of mechanical hibernation, extending over the lunar night with below zero temperatures and had resumed its lunar excursions with all systems functioning normally. This was the first Soviet official report released to the public since the beginning of the lunar night on January 24.

On February 15, Lunokhod-2 began its examination of a large slab formation, apparently thrown from the interior of the moon during the formation of a large crater. Its surface was relatively smooth in contrast to the pock-marked rocks scattered nearby appearing to have been subjected to impacts of meteorites. The investigation of the rocks found at the Taurus Massif indicated that they had remained undistrubed for hundreds of millions of years, whereas those on the slopes appeared to be much younger. Because of the interest in the huge slab, the exploration activities of the rover were restricted for a few days. Its scientific devices were used to study more closely the chemical composition of the slab and investigate the magnetic properties of the moon in that particular area. After a busy period of exploration, Lunokhod-2 was parked again on March 23 for another lunar night lasting until April 8. Following another period of exploration, it was shut down for its fourth lunar night, having traveled 22.5 miles from its lunar base. Its reactivation was not reported by the Soviet space officials after the beginning of the fifth lunar day, starting about May 7.

In its last announcement, released on Monday, June 4, 1973, pertaining to Lunokhod-2, the Soviet space officials stated that the mission had been terminated after four months of lunar exploration. It was maintained that the mission had achieved its objectives, but no clues were given as to why it had been terminated within so short a period. It was suspected by Western space officials that Lunokhod-2 had been involved in some kind of mishap on May 9 as it moved toward the rugged and mountainous terrain.

It was stated that Lunokhod-2 had traveled 37 kilometers (23 miles) during its entire lunar operation. From previous announcements, it was concluded that it had traveled 36.2 kilometers (22.5 miles) by April 22, the time when it became inoperative at the end of the fourth lunar day.[9] On this basis, it would appear that the rover had traveled only a half a mile after being activated at the beginning of the fifth lunar day before the mission was terminated.

It was reported that Lunokhod-2, during its four months of operation involving 60 communication sessions, had transmitted to earth 86 lunar panorama and more than 80,000 television pictures of the moon's surface. It had explored southward from the landing site during the first two lunar days. Then it turned eastward toward a large fault extending ten miles in a north-south manner, where it explored both its western and eastern sides. Along the way, it took pictures and made periodic physical and chemical analyses of the lunar soil and rocks.

On December 17, 1973, Soviet space officials issued the first scientific report of findings of the Lunokhod-2 mission, but no clues were given as to why the mission had been terminated in May.[10] The report contained an account of the scientific experimental accomplishments and was supplemented by a map of 23 miles explored by the lunar rover, following its landing on January 16, 1973. Furthermore, it maintained that the objectives of the mission were to test the soil and rocks and explore features of the lunar surface in the region between the Sea of

Serenity and the Taurus Mountains, south and east of the landing site of Luna XXI. It stated also that laser beams of high intensity sharply-focused lengths were used to track the progress of Lunokhod-2 across the surface of the moon and to measure the distances between the earth station and the moon with high precision.

During the latter part of July and the early part of August, 1973, the Soviet Union again turned its attention to the exploration of the planet Mars. During this brief interval, it launched two sets of Martian-bound spacecraft. At that particular time, the month-long firing window was most favorable for launching spacecraft destined for the vicinity of that planet.[11]

On July 21, 1973, unmanned Mars IV was sent on a mission to Mars with the expectation that it would reach the planet sometime during February, 1974. This marked the first such mission on the part of the Soviet Union in about two years. The purpose of the mission was stated to be the continuation of the scientific exploration of the planet and near-space instituted by the spacecraft Mars II and Mars III missions in May, 1971.[12]

Mars V, the second in this series of unmanned and instrumented interplanetary spacecraft was launched on July 25, 1973. No information was given except that its purpose was to study Mars and its near-space environment. It was implied that the twin mission by identical spacecraft would make it possible to obtain more complete data relating to the planet and physical processes occurring in space. It was reported that on July 25, the two spacecraft were 907,200 miles and 41,010 miles from the earth, respectively.

Again, on August 5, 1973, the Soviet Union hurled Mars VI on its way to Mars. There was speculation that at least one of the trio would attempt to land a capsule on the planet in search of clues as to the existence of some form of life, but the Soviet space officials made no mention of this. The spacecraft did carry a joint Soviet-French experiment known as Stereo, designed to measure solar radiation simultaneously from the spacecraft and ground based stations.

On August 9, the Soviet Union sent Mars VII on its way to Mars, marking the fourth Martian-bound spacecraft launched within a period of three weeks. It was assumed by Western space scientists that this array of spacecraft included both orbiting and landing craft, but again no hint was given as to their specific assignments.

Mars IV and Mars V were identical in design and were launched only four days apart. Mars VI and Mars VII were also identical in design, but different from the first two and were also launched four days apart. It was announced later, that at least one of each pair would perform certain tasks jointly, but no further information was supplied.

During September, 1973, in an unusual break with tradition, a limited disclosure was made by Soviet officials relating to the four spacecraft bound for Mars. It was revealed that they included fly-by-, orbiter-, and lander- type spacecraft. The landers were to test the physical properties of Martian soil and other surface features and conduct experiments with respect to the possible transmission of television images from the planet's surface. No reference was made to any biological equipment aboard the spacecraft, and it was therefore assumed that no attempt would be made to gather information indicative of living organisms on Mars. It was reported also that one of the spacecraft was experiencing some difficulty with its telemetry system, but which one was not identified.

Mars IV failed on February 10, 1974, to assume a Martian orbit and passed the planet at a distance of 1,300 miles because of a malfunction of an onboard system but it did take pictures of the planet with its photo-television instrument. On February 12, Mars V approached the vicinity of Mars and its braking-rocket was successfully fired. It assumed a Martian orbit between 35 degrees north and south of the planet's equator, with a maximum altitude of 18,500 miles and a minimum altitude of 1,000 miles, taking 25 hours to complete. It sent back clear images of the surface of Mars.

On March 14, it was reported that the four-flight mission to Mars had suffered another reverse, this time a double set-back. Radio contact with the capsule ejected from Mars VI was lost before it landed on Mars, but, fortunately, during its descent by parachute, on March 12, some information had been transmitted. It had recorded a major atmospheric component which may have supported some form of living organism and might do so again. It was assumed by some Western space scientists that this component might be argon. It was stated that the space capsule had landed at a point on the Martian surface delineated by 24 degrees southern latitude and 25 degrees western longitude. No indication was given if contact had been restored or could be restored with the capsule after it landed on the surface of Mars. The descent capsule, ejected from Mars VII, shot by the planet missing it by about 800 miles on March 9, because of a malfunction of an onboard system. The spacecraft swung into a Martian orbit as did Mars VI.

The Soviet Union launched unmanned Luna XXII on May 29, 1974. The spacecraft reached the vicinity of the moon and assumed a lunar orbit on June 2, 1974, after 23 radio sessions between it and ground control. The lunar orbit was about 137 miles inclined 19 degrees to the equator of the moon, requiring 2 hours and 10 minutes to complete. Luna XXII did not land on the moon because its mission was to continue studying the moon and its near environment while in orbit, with special attention given to gravitational fields and radiation. It also conducted photographic surveillance of the moon and returned television pictures to earth.[13]

On October 28, 1974, the Soviet Union sent unmanned Luna XXIII into space with the reported purpose of continuing the exploration of the moon and its vicinity. It assumed a lunar orbit on November 2 with an apolune of 65 miles and a perilune of 59 miles, requiring 1 hour and 57 minutes to complete. The spacecraft landed on the moon in the southern part of the Sea of Crises on November 6, but the landing area was unfavorable resulting in damage to the mechanism for gathering samples of lunar rock from a depth up to eight feet. It was made inoperative after a three-day revised testing program without having achieved its mission. This was the first Soviet spacecraft to land on the moon since Luna XXI, which successfully placed Lunokhod-2 on the moon's surface on January 8, 1973.

On June 8, 1975, the Soviet Union sent on a journey to Venus unmanned Venera IX reported to be a new type spacecraft. On June 14, it sent unmanned Venera X, an analogous spacecraft, on its way to the same planet.

Venera IX soft-landed its descent capsule on Venus on October 22, 1975. It was reported that it transmitted information about the cloud cover as it parachuted down to the surface and sent back pictures of the area contiguous to its landing site. A unique picture of the planet's surface was published in the Soviet press which showed rocks of various sizes, some 12 to 16 inches across. One large rock appeared in the distance breaking the smooth skyline. The descent capsule transmitted for 53 minutes from the planet's surface before presumably being destroyed by the intense heat in excess of 900 degrees Fahrenheit and the atmospheric pressure estimated at 90 times that on earth. Venera IX was placed in orbit about Venus *becoming that planet's first artificial satellite.*

On October 25, Venera X landed its descent capsule on Venus, 1,375 miles from the rock-strewn site of Venera IX. It also transmitted information and sent back pictures of the surface of the planet, showing a smoother surface than that shown by the pictures returned by the descent capsule of Venera IX. It transmitted from the surface of Venus for 65 minutes before becoming inoperative. Venera X was placed in a Venusian orbit similar to that of Venera IX.

The data and photographs transmitted by the descent capsules of these Venera spacecraft indicated that the scientists' concepts of the surface of Venus should be reappraised. They appear also to disprove the hypothesis that the horizon of Venus looks like a concave lens.

REFERENCES

Note:

The Soyuz missions, starting with Soyuz X (April 23, 1971) and including those up to and ending with Soyuz XVIII (May 24, 1975), are discussed in Chapter XI because they were primarily related to the space station activities of the Soviet Union. Soyuz XIX (July 15, 1975) was a joint mission with the United States and is dealt with in Chapter XIII as an example of international cooperation. For reference to Soyuz XX (November 17, 1975, which docked with Salyut 4, and Soyuz XXI (July 6, 1976), which docked with Salyut, see Addendum to Chapter XI.

1. It has been estimated that the United States had launched 27 successful civilian and military missions during the same period.
2. *Vostok III Flight,* August 11, 1962, (p. 12).
3. The previous space endurance record of 13 days, 8 hours and 35 minutes was set by *Gemini VII Flight,* December 4, 1965, (p. 84).
4. Mars I was launched on November 2, 1962 (p. 12) and failed when contact was lost with it some 60 million miles from the earth. Zond II was launched toward Mars on November 30, 1964 (p. 13). There is some question whether this mission achieved its purpose.
5. Mars II was launched 11 days after the United States had launched Mariner 8. Mariner 9 was on its way toward Mars.
6. Venera VII launched on August 17, 1970, and landed on Venus, December 15, 1970 (p. 23).
7. It was in this area of the moon that Luna XVI had automatically scooped a sample of lunar soil, and returned it to earth, (p. 23).
8. Luna XVII, launched November 10, 1970, (p. 24).
9. Reported in *The New York Times,* June 4, 1973 (22:1).
10. Reported in *The New York Times,* December 17, 1973, (74:1).
11. Month-long period when Mars is in the most favorable position (nearest to the earth) for launching spacecraft from the earth destined for its vicinity. The firing window occurs every 26 months.
12. Mars II launched May 19, 1971 and Mars III launched May 28, 1971, (p. 25).
13. It was reported by Tass on June 2, 1975 that Luna 22, orbiting the moon for a year as of May 29, 1975, had completed its original flight program and had embarked on additional research. (*The New York Times,* June 3, 1975, 30:8)

Chapter IV

MILLIONS OF POUNDS OF THRUST

United States mastery of space will advance only as rapidly as it can build a fleet of reliable launch vehicles to perform an assortment of tasks—orbit many kinds of scientific payloads and manned capsules, drive spacecraft to the moon, and loft multi-ton interplanetary spaceships to Mars, Venus, and beyond.

NASA (1960)[1]

One of the major factors retarding the early efforts of the United States was the lack of carrier rockets with sufficient power to project heavy payloads into earth orbits.[2] The National Aeronautics and Space Administration (NASA) gave effective guidance to the nation's rocket development programs, and under its direction carrier rockets were used to accumulate weather and scientific data by means of unmanned instrumented satellites, to place men on the moon, return them to earth, and to send spacecraft on interplanetary missions.[3]

Much of the experimentation was carried on through the several branches of the Department of Defense and the National Advisory Committee for Aeronautics, prior to the creation of the National Aeronautics and Space Administration. During the mid-1960's, many of the launch vehicles consisted of components developed in military missile programs, Project Vanguard, and the space program of the United States for the International Geophysical Year. Following this period, the National Aeronautics and Space Administration developed a carrier vehicle fleet based upon standardized units for attaining greater reliability in its series of rockets. It was essential that versatile, inexpensive, and dependable multi-stage carrier vehicles, using solid and/or liquid propellants, be designed and built to carry out its space programs. These programs included placing instrumented capsules in earth orbits at altitudes from 300 to 500 miles and hurling satellites to high altitudes for high velocity reentry tests involving heat and ablation problems. The initial efforts were to develop a series of intermediate-range carrier rockets and a series of long-range carrier rockets.

A successful intermediate-range carrier vehicle development program was the Scout rocket initiated by the National Aeronautics and Space Administration in 1958, with the expectation that it would be fully operational by mid-1960. It was to be a four-stage rocket equipped with a gyroscopic stabilization guidance system with a pre-set program and was designed to test the properties of solid propellants. The controls were to consist of jet vanes and aerodynamic surfaces for the first stage, peroxide reaction jets for the second and third stages, and spin stabilization for the fourth stage. Together, the four stages were to have a length of 70 feet and a total weight of 36,000 pounds: the first stage was to weigh 23,000 pounds with a thrust of 115,000 pounds, the second 8,900 pounds with a thrust of 56,000 pounds, the third 2,200 pounds with a thrust of 13,600 pounds, and the fourth 500 pounds with a thrust of 3,100 pounds. The Scout rocket was to have the capability of placing 200-pound instrumented capsules in 300-mile circular West to East earth orbit or raise 100-pound capsules to an altitude of 6,000 miles.

On April 18, 1960, a dummy Scout rocket was fired to test its third-stage engine and on July 1 of that year, the first guidance and control techniques were used on the four-stage all-solid propellant carrier vehicle. All stages functioned satisfactorily except the fourth, which experienced an ignition failure, causing the combined third and fourth stages to rise to an altitude of 860 miles before falling into the Atlantic Ocean some 1,500 miles from the launching site. Improved solid propellants were developed for use in the third and fourth stages increasing its payload capacity by about forty per cent. In 1962, a new engine, with 23,000 pounds of thrust was installed in its third stage and between July 1, 1964 and June 30, 1966, the Scout rocket was launched 18 times without failure.

Four-Stage Scout Rocket

The Delta rocket was one of the most successful intermediate-range carrier vehicles developed by the United States. The first stage, a modified Thor operational rocket, was liquid-fueled and capable of developing 150,000 pounds of thrust. During its buring period, it opened into four petals after ignition to allow the second stage to separate. The second stage, a modified version of the Vanguard's second liquid-fuel stage (Thor-Able rocket) was equipped with guidance and altitude control systems, together with a steel, instead of an aluminum, thrust chamber. The third stage, an improved spin-stabilized solid-fuel rocket, was equipped with an improved engine, previously used in the Thor-Able and Atlas-Able rockets. The first complete Delta carrier vehicle was delivered to the National Aeronautics and Space Administration in May 1960.

The Delta carrier rocket had the capability of placing 100-to 300-pound instrumented capsules in circular-earth orbits at altitudes of 1,040 miles or into elliptical earth orbits with apogees of 46,100 miles and perigees of 156 miles. It was powerful enough to send a 65-pound payload to the moon. The complete three-stage Delta I was fired for the first time on May 13, 1960 when it thrust the Echo inflatable communication satellite into space. Although the second stage malfunctioned and the third stage did not ignite, the satellite attained an altitude of 1,000 miles.

Delta II, launched on August 12, 1960, placed Echo I a passive communication satellite, into earth orbit. It was designed as an interim rocket not for use after 1961, but was given a new lease on life because of its high reliability and low costs. It became one of the most reliable launch vehicles and as of June 30, 1966, its record stood at 34 successes out of 38 attempts.

The Agena B carrier vehicle, an enlarged and improved version of the Agena A, an Air Force carrier rocket, proved reliable and dependable in the Discoverer Program. It was liquid-fueled, capable of providing 15,000 pounds of thrust and equipped with engines that could be started and stopped in flight. This rocket, the second stage in combination with the 150,000-pound thrust Thor rocket, produced a carrier rocket with a thrust of 165,000 pounds, and with the capability of placing a 1,600-pound payload into a 300-mile earth orbit. The Thor-Agena B carrier rocket was used for launching meteorological, communication, and scientific satellites to be placed in polar orbits about the earth. During January 1964, it put Echo II, a passive communication satellite, into earth orbit and in August of that year, put Nimbus II into polar-earth orbit.

The Atlas-Agena B combination carrier rocket possessed a thrust of 360,000 pounds at lift-off and 80,000 pounds during the sustainer phase of its flight. This multi-stage rocket had the capability of thrusting 800 pounds of payload to impact the surface of the moon or placing a 5,000-pound payload into a 300-mile East-West earth orbit. It was designated originally for use in the unmanned lunar-probing program and in meteorological, communication, and scientific space-probing programs. Later, it was substituted for the Centaur rockets in unmanned satellite launchings to probe Venus and Mars, because of the delay in that rocket's program. On August 23 and November 18, 1961, it was used in two attempts to launch lunar-probing capsules into elliptical orbits of more than 500,000 miles. In both instances, the Atlas performed as planned but the Agena B rockets failed to restart in flight, resulting in failure of the space capsules to enter into solar orbit or impact the moon. On April 23, 1962, a successful lunar impact by Ranger 4 was achieved with the Atlas-Agena B combination. Agena D, a modification of the Agena B, in combination with the Atlas booster rocket, was used as the target during the rendezvous phase of the Gemini program.[4]

Gemini Launch Vehicle (the Titan) with Gemini Spacecraft (left) and Gemini Agena Target Vehicle with Atlas Booster (right) on Launch Pads Shown in Composite Form (Source: Lockheed Missiles and Space Company)

The Atlas-Centaur program was initiated during 1958 to develop a more powerful two-stage carrier vehicle using the Atlas D as the first or booster stage. The Atlas D rocket was to be modified by replacing the tapered forward part of the liquid oxygen tank with a cylinder of the same diameter as the back part of the Atlas rocket. The second stage was to be powered by two turbopump-fed rocket motors each using liquid hydrogen with liquid oxygen as fuels and each having the capability of delivering 15,000 pounds of thrust. The mechanism of the second stage and the payload were to be protected from the heat, caused by friction with the atmosphere, by a cone-shaped shield, to be jettisoned once the danger of the intense heat had passed. This carrier rocket was to have sufficient thrust to place communication satellites into a 22,300-mile equatorial earth orbit with velocities equal to the earth's speed of rotation, causing them to become stationary satellites with respect to the earth's surface. It could be used also to place 8,500-pound payloads into earth orbits and 1,450-pound payloads into lunar orbits.

The first Atlas-Centaur carrier vehicle was fired on May 8, 1962 from Cape Canaveral. After being in flight for 54 seconds, the Centaur exploded in mid-air but the Atlas booster performed satisfactorily. On November 27, 1963, an Atlas-Centaur was fired successfully, putting the five-ton Centaur stage in an earth orbit. *This was the first successful flight of a liquid-hydrogen-fueled space vehicle in the free world.* Another successful launching of this carrier rocket was achieved in June 1964, followed by a partially successful firing on December 11, 1964, when a model of a Surveyor spacecraft was placed into a precise earth orbit. The development program of this direct ascent version[5] of the Atlas Agena Rocket program was completed with its successful lift-off during August 1965.

The seventh launching of the Atlas-Centaur placed its satellite into a planned parking orbit on April 7, 1966, but unfortunately the second burn was not accomplished, because of a leakage in the fuel line. The eighth launch of this carrier rocket put Surveyor I into lunar-intercept trajectory on May 30, 1966, from an earth-parking orbit.

The Saturn launch-vehicle program, started in 1958, was one of compromises between launch rockets of various two-, three-, and four-stage carrier rockets, capable of putting up to ten tons into a low earth orbit. It had a twofold purpose of obtaining rapidly a large payload capability for the United States and at the same time developing acceptable mission reliability through the technique of clustering reaction rocket engines.

The Saturn carrier rockets were the most powerful multi-stage rockets developed in this country but they did not represent the ultimate in size and power capability.[6] The clustering technique[7] made possible the attainment of enormous initial velocity as well as sustaining power. It gave the Saturn carrier rockets the capability of placing into fixed-earth orbits space stations from which spacecraft would be able to take off, land on, and return from the moon and solar planets or circumnavigate them.

The Saturn I launch vehicle, a two-stage rocket, was designed to place payloads of 22,000 pounds into earth orbits. During 1964 seven missions were flown. The eighth launch was on February 16, 1965 and the ninth on May 25, 1965, *was the first night launch in the Saturn series.* The last Saturn I vehicle was launched on July 30, 1965 when it placed a Pegasus satellite into earth orbit, making for ten out of ten successful launchings. This launch vehicle was the stepping stone to the Saturn 1B (uprated Saturn I) and the Saturn V launch vehicles.

The Saturn 1B was a two-stage rocket with sufficient thrust to place the Apollo spacecraft into low earth orbit. It was a compromise of two stages under development in Saturn I and Saturn V projects—the S-I (first stage) and S-IVB (second stage) and was capable of being modified to meet the needs of particular missions. The design of this launch vehicle was completed by the end of June 1965 and testing of subassemblies was completed to about sixty per cent. The first vehicle was successfully flown during the first part of 1966 under the all up principle of testing; that is, all stages were fully operational and committed to flight together in the first test.

The rocket's first stage was similar to the original Saturn I first stage, being fueled with liquid oxygen and kerosene, weighing less and having greater thrust (1,600,000 pounds). Its new second stage S IV-B had a thrust of 200,000 pounds and was liquid hydrogen/oxygen fueled. It was the same as the third stage of the Saturn V.

The Saturn C-I, a two-stage version of the Saturn launch vehicle was modified to conform with the requirements of orbital flights of the missions proposed for the Apollo spacecraft. Fins were added to the surface of the first stage providing greater stability and control. The six 15,000-pound of thrust engines of the second stage were substituted for the four 17,500 pounds of thrust Centaur engines, increasing the total thrust by 20,000 pounds and providing more satisfactory engine-out capability.[8] This modified two-stage vehicle had 1.5 million pounds of thrust for the first stage and 90,000 pounds of thrust for the second stage making it capable of carrying heavier payloads into space. On its first test flight on October 27, 1961, it gave a near-perfect performance, reaching a maximum altitude of 84.8 miles and attaining a top speed of 3,607 miles an hour during an engine burn-time of 115 seconds. It fell into the Atlantic Ocean 214.7 miles down range from Cape Canaveral, 8 minutes and 36 seconds after launch. This flight climaxed three years of effort involving engine clusters and liquid-propelled engines.

The Saturn C-2, a three-stage launch vehicle, was designed to achieve increased thrust. The booster or first-stage engine was capable of providing 1.5 million pounds of thrust, the second-stage engines, 800,000 pounds of thrust, and the third-stage engines, 90,000 pounds of thrust. Even with its tremendous power, it could not launch a payload of the weight of an Apollo capsule into orbits about the moon.

The largest and most powerful carrier vehicle, developed and constructed by the United States to replace the Saturn C-2 booster, was the 36-story Saturn V. It stood 280 feet high, weighed 3,000 tons (six million pounds) and, when fueled, was capable of hurling 240,000 pounds of payloads into space. The combined height of the Saturn V and the Apollo capsule was 365 feet. The first stage was the booster rocket (S-IC)

Atlas-Agena D Launching Lunar Orbiter, August 1, 1967 (Source: NASA)

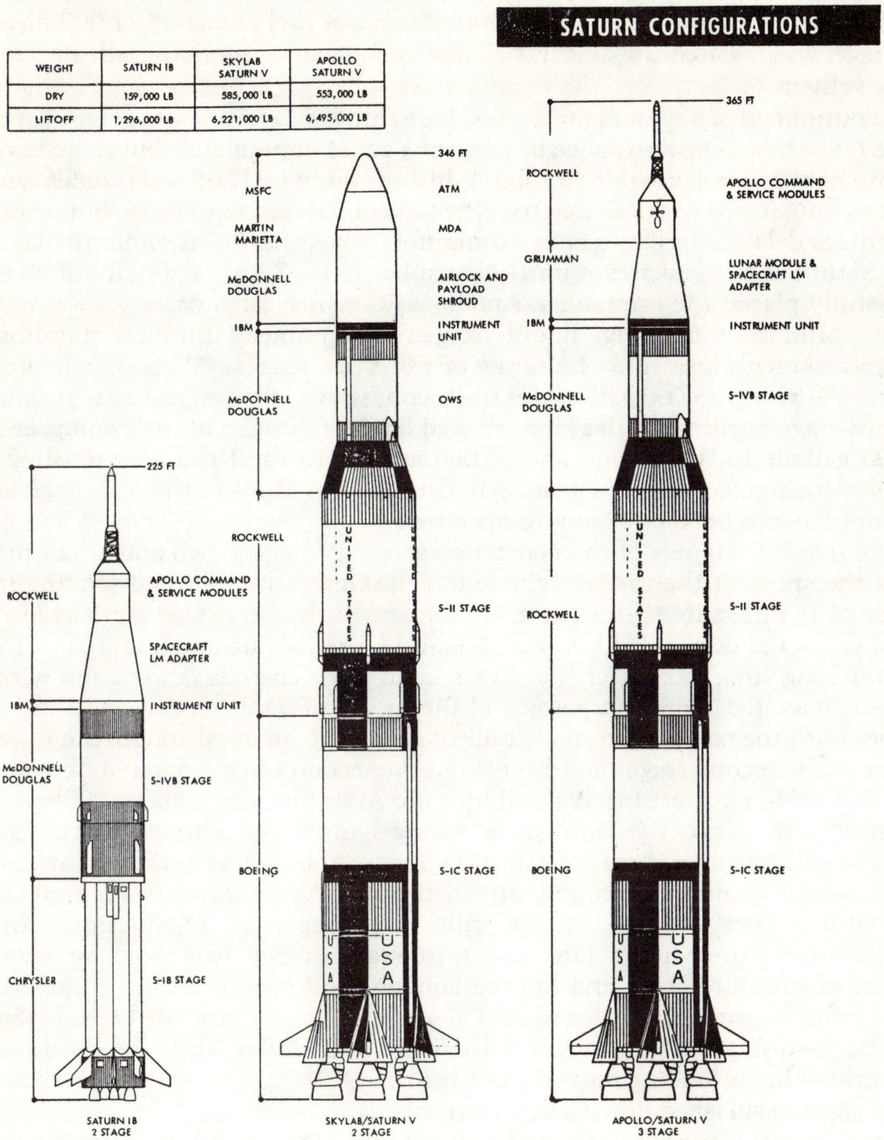

WEIGHT	SATURN IB	SKYLAB SATURN V	APOLLO SATURN V
DRY	159,000 LB	585,000 LB	553,000 LB
LIFTOFF	1,296,000 LB	6,221,000 LB	6,495,000 LB

Saturn Launch Vehicle Configurations (Source: NASA - Photo No. 73-H-254)

which was 33 feet in diameter and 138 feet long. It had a cluster of five F-I rocket engines, each capable of producing 1.5 million pounds of thrust or a combined thrust of 7.5 million pounds for the initial stage of the rocket's flight. The dry weight of this stage was 280,000 pounds and, when fueled, about 4,400,000 pounds. The second stage (S-II) was equipped with five J-2 liquid-fueled engines, each rated at 200,000 pounds of thrust and operating in unison were capable of producing 1,000,000 pounds of thrust. Its diameter was 33 feet, length 88½ feet, and weight, when not loaded for lift-off, 75,000 pounds. The third stage (S-IVB) was equipped with a single liquid-fueled rocket engine as used in the second stage. It had a diameter of 21 feet 8 inches, a length of 59 feet and, without its complement of fuel, weighed 21,000 pounds. In this stage was located an all-inertial guidance system servicing all stages of the carrier vehicle.

A countdown of 83 hours for the test launching of Saturn V was set for September 25, 1967, but had to be postponed because of a series of problems, but it got underway on September 27 for a possible October 1, lift-off, when fuel and test problems delayed the launching further. It was tentatively rescheduled for October 28, but again there was another delay caused by a faulty computer. November 7 was set for the launching of the Saturn V but it was not until November 9, 1967, that it finally lifted off and successfully placed the unmanned Apollo capsule into earth orbit.

Just prior to launching, liquid oxygen was pumped through stainless steel vacuum-jacketed pipes into the tanks of the three stages; 331,000 gallons for the first-stage engines, 83,000 gallons for the second-stage engines, and 20,000 gallons for the third-stage engine. Similar pipes carried liquid hydrogen to the two upper stages, 260,000 gallons to the second and 63,000 gallons to the third. Previously 203,000 gallons of high-grade kerosene had been placed in the tanks of the first stage because it did not have to be kept at low temperature.

The five F-1 engines of the booster stage were fired for two and a half minutes, raising the speed of the carrier vehicle to 6,100 miles an hour and sending it to an altitude of 38 miles. Its engines then cut off, and the booster stage separated from the second stage and descended into the Atlantic Ocean about one hundred miles from the launching site. Soon after, the five second-stage engines ignited and within six minutes generated a million pounds of thrust, forcing the second and third stages, together with the payload, to an altitude of about one hundred and fifteen miles. The engines of the second stage then cut off and the second stage separated from the third stage, and within a short interval fell into the Atlantic Ocean, about 900 miles from the lift-off site. Next, the third-stage engine ignited and burned for two minutes providing an additional thrust of 250,000 pounds. About eleven minutes after lift-off, the third-stage engine cut off and, attached to the Apollo capsule, entered an earth orbit with a peak altitude of 119 miles traveling near 17,500 miles an hour. Approximately three hours later the third-stage engine was fired for about five minutes as the third stage and the unmanned Apollo capsule were completing the second orbit, boosting the speed to 23,370 miles an hour. The space capsule separated from the third stage and rose to an altitude of 11,286 miles while the third stage and the dummy lunar-landing module fell into the Pacific Ocean, shortly before the Apollo capsule splashed down and was recovered.

The ultimate in rocket power has not been attained and there are in the conceptual and developmental stages larger and more powerful carrier vehicles with thrust capabilities far in excess of those now operative. The proposed Nova multicarrier vehicle,[9] if developed, would employ clusters of eight F-1 engines, each engine being capable of producing 1.5 million pounds of thrust in its first stage and clusters of four M-1 engines, each capable of developing a million pounds of thrust in its second stage. Its third stage with a single engine would be similar to that of the advanced Saturn. Ultimately with the high-energy fuels and improved nozzles, Nova should attain a thrust of about twenty million pounds. It would be used to place into space earth-orbiting spacecraft, weighing in excess of 185 tons; lunar spacecraft weighing 75 tons; and interplanetary spacecraft, weighing 50 tons. With a nuclear upper stage, its escape payload would be doubled.

Other sources of energy are being investigated.[10] Nuclear reaction as a replacement for chemical combustion is being experimented with by space scientists. Nuclear reaction could be used to heat propellants which, expanding in combustion chambers, would exert tremendous pressure as they were exhausted through nozzles. Another possible source of energy is based on the acceleration of ions in a controlled direction by an electrostatic field. The National Aeronautics and Space Administration is studying also electrical propulsion systems for rockets to provide propulsion requirements for high-energy-consuming and long-duration space missions.

REFERENCES

1. NASA, *5th Semiannual Report* to Congress, October 1, 1960-June 30, 1961, p. 60.
2. The basic information underlying this chapter was obtained primarily from the *Semiannual Reports of the National Aeronautics and Space Administration* (NASA).
3. Public Law 85-568, 85th Congress, National Aeronautics Act of 1958, July 29, 1958. Under this law the civilian space programs were made the responsibility of the National Aeronautics and Space Administration.
4. See pp. 85-88.
5. There are two basic methods of placing satellites into orbit about distant planets or impacting their surfaces. One utilizes space stations or parking satellites in earth orbit from which space capsules are launched further into space. The second, the "direct ascent" method, relies on large multi-stage launch vehicles to hurl a combination of excursion modules, service modules, and command modules on interplanetary journeys.
6. See p. 40.
7. The development of tremendous thrust is dependent on the development of powerful rocket motors, such as the 1.5 million-pound thrust single-chamber rocket engines. These engines, when arranged in clusters, are able to impart enormous thrust to booster rockets and other stages of rockets where multi-engine arrangements are employed.
8. If one engine should fail to ignite, the other engines would consume its propellants without lessening significantly flight performance.
9. *Nasa 6th Semiannual Report to Congress,* July 1 - December 31, 1961, p. 88.
10. *Nasa 15th Semiannual Report to Congress,* January 1 - June 30, 1966, p. 113. See: Arthur C. Clarke, *Interplanetary Flight,* (New York: Harper & Brothers, Publishers, 1951) p. 90.

Chapter V

THE PROBING OF THE MOON, VENUS AND MARS BY UNMANNED SPACECRAFT OF THE UNITED STATES

> *So that notwithstanding all these seeming Impossibilities, 'tis likely enough that there may be a Means invented of Journeying to the Moon; and how happy shall they be that are first in this Attempt.*
>
> *John Wilkins, Lord Bishop*[1]
> *of Chester (1638)*

Man's insatiable curiosity drives him to investigate the unknown and supplies the basic motivation underlying space exploration and travel. There are scientific reasons and, if useful resources are found on other planets, there may be economic reasons for exploiting and colonizing some of them. Prestigious as well as military reasons may underlie the desire of a nation to be first in achieving spectacular accomplishments in space.

Early conceptions of celestial bodies were based on theoretical deductions derived from numerous findings of telescopic observations, scientific data, and mathematical computations. Today, these theories are being verified or disproved through space exploration by unmanned and manned missions.

Many of the nation's early space-probing missions were directed toward the moon, Venus, and Mars, the greatest emphasis being placed on the moon.[2] These missions have been expanded to include those involving the exploration of Jupiter and Mercury with increased attention being given to unmanned missions to Saturn. Grand tours of planets, by unmanned spacecraft to Uranus, Neptune, Pluto, and beyond the solar system, have been proposed and may come to pass before the end of the century. While these grandiose missions are being planned, manned earth-orbital, moon-orbital, and moon-landing missions have been accomplished in man's search for a better understanding of the universe.

The primary objectives of the moon-probing environment were to obtain information relating to the near-lunar environment and the nature and physical characteristics of the lunar surface. It was pertinent to collect and analyze extensive data relating to its environment and surface before sending manned spacecraft to land on its surface. Unmanned exploratory missions were carried out in three phases:

crash-landing of instrumented capsules on the moon, soft-landing of them on its surface, and orbiting instrumented capsules about it.

On January 31, 1958, following the Soviet Union's success in placing Sputnik I into earth orbit, the United States hurled its first space capsule, Explorer I into earth orbit with an apogee of 1,590 miles and a perigee of 220 miles, inclined at 34 degrees to the earth equator, requiring 115 minutes to complete. The capsule was cylindrical in shape, measured 80 x 6 inches, weighed 30.8 pounds, carried a payload of 11 pounds, and attained a speed of 18,000 miles an hour. The expected life of the capsule was five years.[3] This marked, for practical purposes, the inauguration of the phase of the nation's space mission dealing with the probing of the moon and solar planets with unmanned spacecraft. During the last three months of 1958 and in March 1959, the United States launched four unmanned moon-probing missions of the Pioneer series, of which three were partially successful and one a failure.

Pioneer 1 was launched on October 10, 1958 from Cape Canaveral (now Cape Kennedy).[4] The capsule weighed 39 pounds and measured 30 inches in length and 29 inches in diameter. It encountered difficulties after lift-off, even though the first and second stages of the carrier rocket functioned satisfactorily. The third stage burnout velocity was approximately five hundred feet per second less than needed to escape the earth's gravity, causing the spacecraft to yaw 16 degrees and pitch 15 degrees. An attempt was made to convert the mission into a high altitude flight by firing the fourth-stage rocket but it could not be ignited because the interior temperature of the rocket was not high enough for the mercury batteries to ignite. The spacecraft traveled 70,700 statute miles in space above the earth on a journey that lasted 43 hours before it reentered the earth's atmosphere and disintegrated.

Pioneer 2, having as its goal the same objectives as Pioneer 1, was sent into space on November 8, 1958. It blasted off on schedule with the first and second stages of the carrier rocket functioning satisfactorily. The third stage failed to ignite, but it and the fourth stage rose to 963 miles above the earth and traveled 7,500 miles before reentering the earth's atmosphere and disintegrating.

On December 6, 1958, Pioneer 3 was launched carrying an instrumented and gold-plated capsule weighing 12.95 pounds. The mission was a qualified success. Its primary purpose was to carry out scientific exploration in the near-lunar environment but this was not accomplished because the first stage of the carrier rocket cut off several seconds prematurely. This prevented the spacecraft from attaining the necessary velocity to escape the earth's gravitational force. The spacecraft traveled 63,580 miles in space during a period of 38.6 hours of flight before reentry and disintegration over Africa. However, useful information was obtained relating to the operational characteristics of the carrier rocket and the steps which should be taken to avoid similar malfunctions.

During the flight, the magnetometer, micrometeorite detector, command receiver and transmitter functioned satisfactorily, but the ionization chamber measuring-indicator developed a leak making it difficult to interpret information being collected. Nevertheless, worthwhile information was obtained verifying the existence of an inner radiation belt extending to about 2,000 miles and an outer reaching out to about 10,000 miles from the earth. Beyond 10,000 miles radiation dropped rapidly and became exceedingly weak after 40,000 miles. Unfortunately the capsule did not approach the moon closely enough to activate its scanner.

The Pioneer 4 mission, initiated on March 3, 1959, was more successful. In this instance, the capsule was constructed of gold-plated fiber glass weighing 13.40

pounds and measuring 20 inches in length and 9 inches in diameter. Its instruments included a battery-powered radio, two Geiger-Mueller tubes for measuring radiation, and a photo-electric radiation sensor activated when the capsule came within 20,000 miles of the moon.

The primary objectives of the mission were to achieve a suitable earth-moon trajectory, obtain radiation data, and provide experience in tracking space vehicles. The spacecraft did not approach the moon closely enough to have its radiation scanning sensor activated but sufficient information was obtained to fulfil its other objectives. On March 4, 1959, 41 hours and 13 minutes after lift-off, the spacecraft passed within 37,300 miles of the moon. Up to March 13, the ground control stations were able to track it for 82 hours and 4 minutes covering a distance of 407,000 miles from the earth, *the greatest distance that a man-made satellite had been tracked up to that time.* Pioneer 4, after flying by the moon, assumed a solar orbit reaching its perihelion 91.7 million miles on March 17, 1959 and its aphelion, 106.1 million miles on October 1, 1959.

On March 11, 1960, Pioneer 5 was hurled into space on a solar-orbiting mission with a velocity of 24,886 miles an hour. It was launched counter to the earth's revolution in order to attain this velocity, which was 575 miles an hour faster than the minimum speed required to escape the earth's gravitational force and place it in a solar orbit. The 26-inch spherical capsule with four solar vanes, weighed 94.8 pounds and was one of the most elaborately instrumented satellites ever designed to accumulate scientific data and test its transmission from deep space. *This was the first time that an instrumented capsule had been sent into the depth of the solar system to obtain and relay back to earth information relating to the role played by the sun in our lives.* The data-gathering capability of the space capsule was restricted to the range of its radio transmitter of 50,000,000 miles in space. This was expected to be adequate to cover its nearest approach to the sun during its first solar orbit, then it would pass out of its radio transmitter's range and contact with it would be lost. The capsule also carried devices for detecting x-rays and ultraviolet radiation.

It followed a trajectory taking it between the earth and Venus. During its first solar orbit, it was estimated that it would be traveling at an average of 67,750 miles an hour as compared with the earth's average orbital speed of 66,593 miles an hour and that of Venus of 78,403 miles an hour. In its solar orbit, its closest approach to the sun was 74,967,000 miles reached on August 10, 1960, and its farthest distance from the sun was 92,358,000 reached on January 13, 1961. The greatest distance it would recede eventually from the earth would be about 186 million miles, and the closest it would come to the earth in the subsequent decade would be probably several hundred thousand miles. It might even be 100,000 years before it would approach closely enough to disintegrate in the earth's atmosphere. Each circuit of the sun was equal to a total of 514,500,000 miles requiring 311.6 days to complete.

The radio transmission test was one of the five principal scientific experiments carried on board Pioneer 5. The four others were designed to measure: high-energy radiation in space, particularly that emitted by the sun during solar disturbances; total radiation, encountered in the interplanetary space; magnetic fields in space; and the density of micro-meteorites or cosmic dust.

On March 13, radio signals were received from Pioneer 5 revealing that it had passed beyond the 407,000 mile mark set by Pioneer 4, establishing a new record for radio transmission in deep space. On March 18, its five-watt transmitter sent data from one million miles in space on a command from its ground tracking station at Kaena Point, Hawaii. The data consisted of cosmic ray counts, a record of micrometerorite impacts, temperatures inside the space capsule and on its surface, and magnetometer measurements.

Ranger 1 and 2 were launched during 1961, the former on August 23 and the latter on November 18.[5] They were six-sided capsules with two folding solar panels, designed to provide power for operating their radios and other instruments. In both instances, malfunctions, in the second-ignition phase, caused the satellites to enter low-flight orbits at altitudes between 312 and 95 miles. Ranger I remained in earth orbit for about a week and Ranger 2 about six hours, before reentering the earth's atmosphere and disintegrating. The missions were designed to obtain information about the surface of the moon and to land survivable instruments as well as test the spacecraft.

Ranger 3 was lifted into space by the carrier vehicle Atlas-Agena B on January 26, 1962. It provided too much boost, causing it to miss the moon by some 23,000 miles and go into a solar orbit. Although it failed to rough-land on the moon, it did provide valuable information relating to altitude control and locking on the sun as the means of navigational control. It demonstrated also that complex operations could be performed while a spacecraft was in an earth-lunar trajectory. While in flight, the spacecraft responded to command signals from the ground to execute a midcourse maneuver, reorient itself, and acquire a sun-earth alignment. Its gamma-ray spectometer, operating for 50 hours, sent back information relating to the space environment passed through on its way to the moon.

On April 23, 1962, Ranger 4 was fired into an earth-moon trajectory and 64 hours later crash-landed upon the moon's surface, *and was the first spacecraft of the United States to do this.* Unfortunately, it was damaged on impact, and was unable to transmit the information to its ground control station. However, its seismometer continued to send out signals making it possible to track it up to the time it impacted the moon's surface.

The exploration of space by the United States was not confined solely to the moon. Under the Mariner Project,[6] unmanned instrumented spacecraft were sent near Venus and Mars, carrying microwave radiometers to measure their temperatures; ultraviolet spectrographs to examine their upper atmospheres and search for water vapor and other components; neutron counters to measure the ratio of carbon dioxide to nitrogen in their atmospheres; magnetometers to determine the approximate strength and character of their magnetic fields; and scintillation counters to sample radiation trapped in their magnetic fields. By means of these unmanned scientific exploratory missions, data were accumulated which may provide answers to vexing questions relating to the probability of the existence of some form of living organisms on these planets.

As early as 1962, an attempt was made to send an unmanned spacecraft to the vicinity of Venus. On July 22, 1962, Mariner 1 was launched on a fly-by mission to that planet but went out of control about five minutes after lift-off, causing the mission to be terminated deliberately.

On August 27, 1962, Mariner 2 was fired into a trajectory taking it past Venus at a distance of 21,648 miles on December 14. *This spacecraft established a record for long-distance communication when, on January 3, 1963, it transmitted information over a distance of 53.9 million miles before ground stations lost contact with it.* During its four months in space, it transmitted 90 million scientific data points, indicating that Venus had a surface temperature of 800 degrees (Fahrenheit), and −30 degrees (Fahrenheit) within its cloud cover. The data provided no evidence of magnetic fields along the path of the spacecraft nor of trapped particles in radiation belts. From preliminary information, it was concluded that the mass of Venus was 0.815 of that of the earth.

Ranger 5 was launched on October 18, 1962, on a lunar mission to gather information about the surface of the moon and its environment. The 727-pound spacecraft, with a survivable instrumented capsule, did not achieve its goal, because approximately thirty-eight minutes after injection into lunar trajectory from an earth-parking orbit, its solar panel malfunctioned, forcing it to rely solely upon its conventional batteries for its electrical energy. Contact with it was lost after about ten hours in flight. Because no midcourse maneuver could be executed, it could not land a survivable capsule on the moon and went into a solar orbit. This failure led to a review of the total Ranger Program by a Board of Enquiry.

Ranger 6 (A), one of a new series of lunar-probing satellites, was launched on January 30, 1964 and crash-landed on the surface of the moon on February 2, 1964. The launching was so accurate that the spacecraft impacted the moon's surface within one second of the calculated time of impact and within twenty miles of the planned target. The mission failed, however, to accomplish its photographic surveillance of the moon, in spite of the fact that it was equipped with six television cameras capable of taking 3,000 lunar pictures and transmitting them to earth during the ten-minute interval prior to impact. It may have been that a premature activation of the television subsystem during the launch phase caused the failure of the mission.

Ranger 7 (B), with modifications, was launched on July 28, 1964 and followed a moon trajectory causing it to impact the moon only seven miles from the planned impact point. Approximately seventeen minutes before crash-landing, its full-scan television cameras began transmitting pictures to earth, followed five minutes later

Mariner 2 Spacecraft (Source: NASA - Photo No. P-1963)

by its partial-scan television cameras. The six cameras together transmitted 4,316 pictures of the lunar surface during that period. Those taken just before impact, showed surface features as small as 15 inches that were 2,000 times clearer than similar observations made with powerful earth-based telescopes.

Mariner 3, aimed at a fly-by of Mars, was launched by an Atlas-Agena carrier vehicle on November 5, 1964. Unfortunately, it was unable to complete its mission because its shroud failed to separate from it in flight.

On November 28, 1964, Mariner 4 was launched on a trajectory for a fly-by of Mars. In the course of its journey, when *over 109,000,000 miles from earth, it established a new long-distance transmission record.* On July 14, 1965, it flew within 6,118 miles of Mars and took 22 high-quality pictures of the planet's heavily cratered surface. Although the pictures covered less than one per cent of the planet's surface, they revealed differences in altitudes as great as 13,000 feet within a single frame. The visible surface appeared to be extremely old and no arid ocean features were evident. No radiation belts or magnetic fields were detected by sensors and the atmospheric surface pressure was found to be 0.5 per cent of that at the surface of the earth.

Mariner 4 went into lunar orbit after passing by Mars, and continued to transmit signals. Those received from it in January, 1966, when it was more than 216,000,000 miles from the earth, *established another record for long-distance transmission of one-way signals.* During March, 1966, signals were received from it again as it came from behind the sun, *marking the first time that such signals had ever been received*

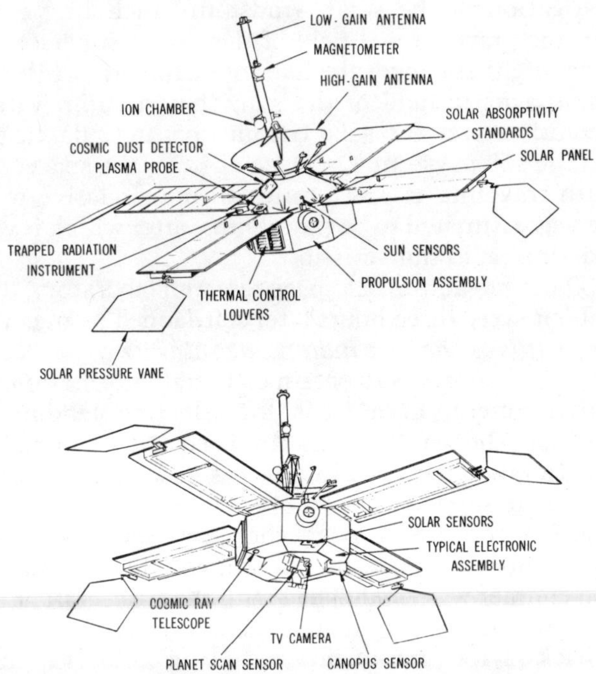

Mariner 4 Mars Spacecraft Identifying Instrumentation

from a spacecraft after it had traveled deep within the solar corona. As of June 30, 1967, Mariner 4 was 57.5 million miles from the earth and 136 million miles from the sun. Telemetry data received during December 1967 revealed that its attitude control gas was exhausted, causing it to oscillate about its three axes. Tracking of Mariner 4 was terminated on December 20, 1967, but during its transmission life, it had covered more than 1.5 billion miles in space and had sent back valuable information concerning the Martian environment and interplanetary space.

Ranger 8 was launched on a lunar mission on February 17, 1965 and crash-landed on the Sea of Tranquility on February 20, within 23 seconds of the predicted time of impact and within 15 miles of the selected impact site. Prior to crash-landing, its television cameras transmitted to earth over 7,100 pictures of the lunar surface, showing the marginal highlands of the Sea of Tranquility and the moon's surface near the point-of-impact. Objects of less than 30 inches in size were shown in the photographs taken just prior to its crash-landing on the moon.

The last spacecraft in the series, Ranger 9, was launched on March 21, 1965 and on March 24, crashed on the moon's surface as planned. The selected target area was in the Alphonsus Crater, of scientific interest because gases appeared to be rising from it. The ground station executed a terminal maneuver in order to direct the spacecraft's television cameras in the direction of the spacecraft's forward movement so that each picture would show the floor of the crater. As it approached the moon, live television pictures of its surface were viewed by millions on earth. Some of the 5,814 high-resolution pictures taken on this mission revealed objects as small as 12 inches.

On December 16, 1965, the 140-pound unmanned interplanetary drum-shaped spacecraft, Pioneer 6, was hurled into a solar orbit.[7] The mission had as its major purposes the investigation of the solar windstorm, including electrically charged gases escaping from the gravitational field of the sun at supersonic speeds, and the extent to which they might endanger the lives of astronauts while in space. It was to investigate also the magnetic field of the sun, chart it from various earth-orbital positions, and differentiate between solar cosmic rays and galactic cosmic rays originating from beyond the solar system. On December 26, the spacecraft was 1,241,039 miles from the earth traveling at 3,912 miles an hour relative to the speed of the earth. Its useful life was estimated to be six months, after which it would pass beyond the range of the existing radio transmission.

Surveyor 1, a 2,200-pound, automatic space research laboratory, lifted off its pad on May 30, 1966 and about sixty-three hours later soft-landed by means of its three legs, on the Sea of Storms.[8] *It was the first man-made satellite to soft-land as planned on a celestial satellite.* Its mission was to obtain essential information about the moon's surface and its environment, as a basis for selecting landing sites for future manned lunar missions. The satellite was equipped with four cameras and a high-resolution telescope to take pictures while it was descending toward the surface of the moon, scan the lunar surface after landing, and monitor its instruments while they were operating. The concern as to whether Surveyor 1's systems would survive the extended period of intense cold of the lunar night (14 earth-days) proved groundless. On July 6, 1966 contact was established with the spacecraft at the end of its first lunar night and on the following day its television cameras were activated and took 24 pictures. It sent back 11,150 high-quality television pictures of the light side of the moon, revealing that its surface was firm enough to support the weight of the Apollo spacecraft being designed for manned lunar-landing missions.

Full-Scale Mock-up of Surveyor Lunar Lander Spacecraft (Source: NASA - Photo No. 66-H-476)

Lunar Orbiter 1 was launched on August 10, 1966 from Cape Kennedy on a 92-hour journey and assumed a lunar orbit on August 14 with an apolune of 1,048 miles and a perilune of 130 miles. Later, it entered an orbit with an apolune of 1,100 miles and a perilune of 36 miles on its photographic reconnaissance of the moon.[9] *It was the first American spacecraft to accomplish this feat.* Its primary purpose was to photograph a 3,000-mile strip along the equator of the moon containing nine possible landing sites for the spacecraft of the planned Apollo missions. It was also to photograph what remained of Surveyor 1, soft-landed on the moon on June 2, 1966. By September 15, 1966, Lunar Orbiter 1 had completed its photographic mission, having taken 215 pictures *among which were the first photographs of the earth from the vicinity of the moon.* It continued to orbit the moon until October 29, 1966, when it was deliberately crash-landed on the surface of the moon to prevent its signals from interfering with those of Lunar Orbiter 2 to be launched soon.

The 140-pound drum-shaped sun-orbital spacecraft, Pioneer 7, was launched from Cape Kennedy on August 17, 1966, and by mid-afternoon was 40,000 miles from the earth with all systems functioning satisfactorily. It was anticipated that the spacecraft would take twenty-eight weeks to cover the 12 million-mile journey beyond the earth's orbit and that its closest approach to the sun would be 94 million miles. It was successfully placed in a solar orbit with an aphelion of 105 million miles and a perihelion of 94 million miles requiring 403 days to complete. The primary purpose of the mission was to study the supersonic wind and the lethal cosmic rays emitted by the sun. The assembled information was to be used for forecasting the solar weather and was to be coordinated with that obtained by other space flights already in progress.[10]

Surveyor 2 was sent into space on September 20, 1966 but failed to achieve its mission. A malfunction of the carrier rocket during the midcourse maneuver caused it to crash-land rather than soft-land on the surface of the moon.

An Atlas-Agena carrier vehicle put the 850-pound windmill shaped Lunar Orbiter 2, a flying photographic laboratory, into space on November 6, 1966 and placed it into a lunar orbit. The orbiter was constructed to process onboard and transmit to earth more than 400 pictures during an eight-day period. Its cameras had 24-inch lenses and was designed to take high-resolution and medium-resolution 35 mm. pictures simultaneously. The high-resolution pictures were to be taken when the spacecraft was about twenty-eight miles above the surface of the moon, showing small objects 40 inches in width, and in medium-resolution pictures taken at higher altitude, showing objects 25 feet in width. The spacecraft ended its photographic mission unexpectedly on December 8, when it failed to respond to signals from earth, but by that time it had accomplished about ninety-seven per cent of its mission. In all, it had relayed to earth 97 per cent of the 211 exposed frames covering more than 15,000 square miles of the moon's surface along the lunar equator.

On February 4, 1967, Lunar Orbiter 3 was launched on a site-confirmation mission to the vicinity of the moon and entered an initial lunar orbit with an apolune of 1,118 miles and a perilune of 131 miles with an inclination of 21 degrees to the moon's equator, requiring 3 hours and 35 minutes to complete. Before undertaking its lunar photographic activities, its retrorocket engines were fired bringing it within about 30 miles of the moon's surface, from which height it took 360 pictures covering the eastern edge of the Sea of Tranquility along the moon's equator near its eastern edge. The pictures showed that this area was pocked with small craters, some with diameters of three feet and others 60 to 70 feet, and one or two with diameters of the craters approximating to four hundred feet. Other features shown were ridges, some steep enough to overturn a landing spacecraft, a huge fault estimated to be four hundred miles long and as deep as the Grand Canyon, and a 6,500-foot mountain rising from the floor of a large crater.

Success accompanied the launching of Surveyor 3 on April 17, 1967. It was hurled into space by a combination Atlas-Centaur rocket and after a near-perfect flight of 65 hours, landed within 2.4 miles of its planned lunar target, southeast of the Landsberg crater on the eastern edge of the Ocean of Storms. The landing site was about three hundred and eighty miles east of where Surveyor I had landed on June 2, 1966. Surveyor 3's mission was to continue tests of the lunar soil with the aid of a television camera and a robot arm with a claw which could be extended about thirty inches. One of the tests of the capability of the moon's surface to sustain weight, required the arm to be lowered and pressed against the moon's surface under eight pounds of pressure for five minutes making a trench one and a half inches deep. By April 22, it had sent back about 1,500 pictures, the first few of which were viewed on earth by television in millions of American homes.

Lunar Orbiter 4 was launched from Cape Kennedy on May 4, 1967, and went into a lunar orbit with an apolune of 3,844 miles and a perilune of 1,623 miles. It had as its primary purpose, the exploration of the moon's surface by photography. This mission differed from the previous Lunar Orbiter missions in that it assumed a higher orbit enabling it to take more distant pictures of selected areas of the moon's surface and cover 80 to 90 per cent of the front and rear hemispheres of the moon. The initial pictures were transmitted on May 11, *including for the first time views of the moon's South Pole.* The broad panorama of the rugged moonscape, from an altitude of 2,176 miles, showed high ridges, many straight-line ridges, and numerous craters of various sizes, their dimensions being indicated by the varying lengths of shadows. The photographic mission of Lunar Orbiter 4 was terminated on May 26, because of a malfunction of a switch, but up to that time, it had taken 163 of the planned 180 pictures. It continued to transmit information until July 25, when radio contact was lost. On January 31, 1968, it crash-landed on the lunar surface, on command by ground control.

Mariner 5, a 542-pound spacecraft, was fired into space on a 212.5-million mile journey to the vicinity of Venus on June 14, 1967, estimated to take about four months. This launching occurred two days after the Soviet Union sent the 2,437-pound Venera IV toward the same planet. Except for changes in onboard scientific instruments, it was similar to Mariner 2, which approached to within 21,648 miles of Venus in 1962, and Mariner 4, which took the first close-up photographs of the planet in 1965.

Scale Model of Lunar Orbiter and of Lunar Surface. The Lunar Orbiter was planned to photograph the lunar surface for selection of landing sites. (Source: NASA-Photo No. 66-H-435)

On October 19, 1967, Mariner 5 flew within 2,480 miles of Venus, traveling at 19,122 miles an hour, curved behind it, and continued on into space. It was anticipated that it would approach to within 54,000,000 miles of the sun by January, 1968. Its sensing device was set to detect the light from the planet and automatically turn on a tape recorder, equipped with a 50-foot loop magnetic tape. Clear electronic impulses, requiring about forty thousands words to interpret (six digits represented a data word), were recorded on the tape and were later played back to earth by its small onboard ten-watt radio, taking 34 hours to accomplish. Mariner 5's fly-by took place about thirty-six hours after the Soviet Union's Venera IV ejected an instrumented capsule to land on Venus. The spacecraft completed its basic mission on October 21, 1967, having recorded valuable data about Venus transmitted to earth from a distance of 50 million miles in space. It continued to orbit the sun and would again come within communication range of the earth in August 1968.

The conclusions based on a preliminary study of the data were that Venus had: no radiation belt; a definite magnetic activity 1/300 as strong as that of the earth; an atmosphere beginning at 3,800 miles from the planet's center, composed largely of carbon dioxide (75 to 85 per cent) with traces of hydrogen but no detectable oxygen; an emission of a faint ultraviolet glow from its night side; temperatures in its upper atmosphere of about 700 degrees (Fahrenheit) above zero and about 500 degrees above (Fahrenheit) on its surface; and a halo, a reflection of sunlight on hydrogen atoms.

Surveyor 4, carried into space by an Atlas-Centaur rocket on July 14, 1967, after being postponed for a day because of a short-circuit in the carrier rocket, failed in its

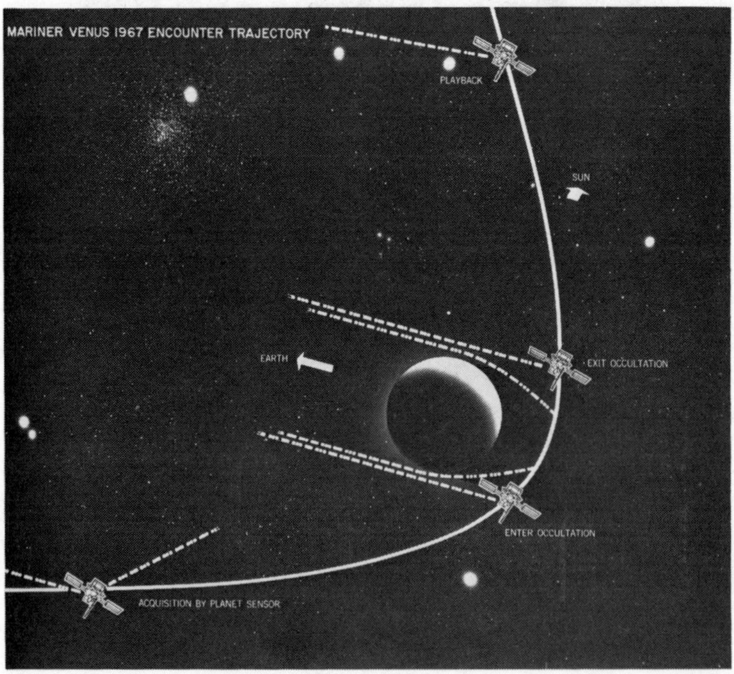

Mariner Venus Flyby Trajectory 1967

mission. Its moon trajectory was so accurate that it would have reached the moon within about 125 miles of the intended 37-mile wide landing site in Sinus Medii. After a successful lift-off and a promising flight, ground stations lost contact with the spacecraft on July 16, about three minutes before it was to land and it was concluded that it had crash-landed on the moon. It was equipped with a camera, a magnet, and a shovel for testing the lunar soil. The shovel was a miniature device capable of digging trenches in the lunar soil and dumping it on one of the capsule's legs containing a two-inch bar magnet for detecting its possible iron content.

The last spacecraft of the project, Lunar Orbiter 5, was launched from Cape Kennedy on August 1, 1967 for the primary purpose of completing the examination of five potential sites for the planned Apollo project. In addition, it was to investigate the terraced Copernicus crater; the hot spots of the Aristarchus Crater; the Cobra Hind Valley (inside the Schroeter Valley); the volcanic region known as the Harbinger Mountain; and other deep depressions. The spacecraft was in an eliptical orbit with an apolune of 3,740 miles and a perilune of 122 miles when its camera was activated. It later dropped into a lower orbit with an apolune of 125 miles and a perilune of 60 miles for its planned 14-day photographic mission. It sent back pictures of a heavily cratered area on the backside of the moon, followed later by photographs of possible landing sites on the front side. On January 31, 1968, it was ordered to destroy itself on signals from ground control causing it to crash on the moon's surface because of depleted fuel supply.

Surveyor 5 lifted off on September 8, 1967, but its rocket developed a leak in the fuel system while in flight. It was debated whether to terminate the moon mission and put the spacecraft into an earth orbit or let it continue on its moon trajectory for almost certain destruction by crashing on the moon's surface. Finally, it was decided to allow it to continue hoping that it would soft-land on the moon. This it did on September 10, 1967, coming to rest on the Sea of Tranquility about two miles from the planned landing site.

It carried a rotating camera to search for landing sites for the planned Apollo missions and was *the first lunar landing equipped to test the chemical composition of the lunar soil.* The basic instrument for conducting the soil analysis was a six-inch square metal box, suspended by a nylon cord from the spacecraft, containing a radio-active sensor for bombarding the soil with atomic particles and measuring their rebounds. Within 1 hour and 15 minutes after landing, Surveyor 5 transmitted radio signals and began transmitting television pictures of its superstructure and the relatively level moonscape. By November 1967, it had transmitted 18,006 photographs of the moon. Two unsuccessful attempts were made to reactivate the Surveyor after the end of the lunar night.

Surveyor 6, launched on November 7, 1967, was also equipped with a television camera and instruments for testing the lunar soil. It soft-landed on the moon's surface within three miles of the planned target in the Central Bay or Sinus Medii, about five miles from where it was believed that Surveyor 4 had crash-landed. The transmitted pictures were of a high quality and among the best close-views of the moon's surface received up to that time, showing physical features which appeared to be cliffs, trenches, and craters. By November 11, it had sent back 2,300 photographs.

The missions for 1967, were terminated with the launching on December 13, of the Pioneer 8 spacecraft involving a two-in-one shot. At the same time that Pioneer 8 blasted off to enter a solar orbit, with an aphelion of 101 million miles and a perihelion of 92 million miles, a small communication satellite was also launched by

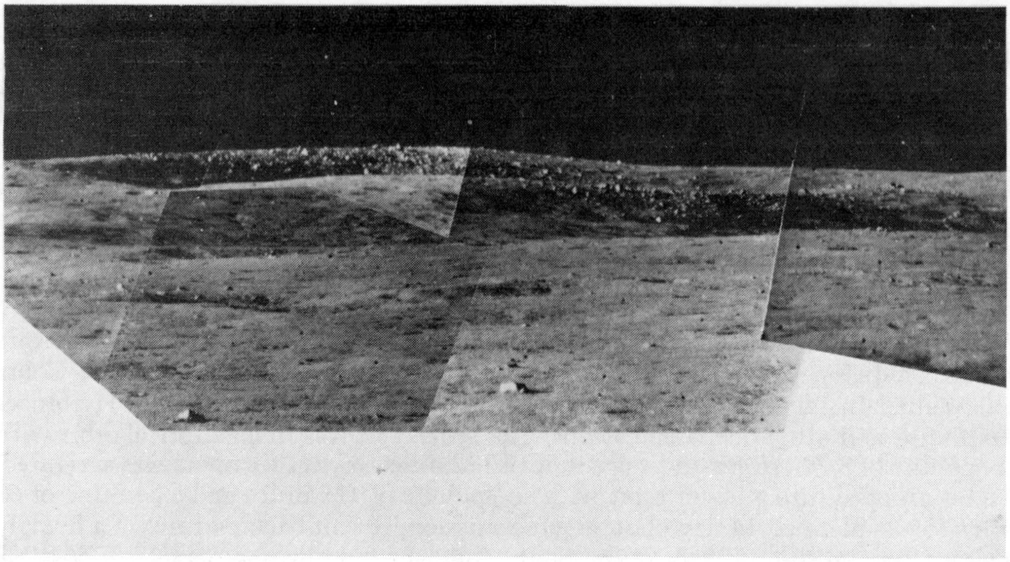

A Five-Frame Mosaic of the Lunar Surface near Surveyor 6. Some rocks are as large as 1.5 to 2 ft. across and are believed to have been thrown out of a crater beyond the horizon. (Source: NASA - Photo No. 67-H-1551)

the same rocket to enter an earth orbit. *This was the first time that a single rocket of the United States had been used to put two payloads into two widely separated areas in space simultaneously.* Both satellites were associated with plans for space flights to the moon and planets of the solar system. Pioneer 8's mission had goals similar to those of Pioneer 6 and Pioneer 7, and as of January 2, 1968, was orbiting about the sun with all onboard systems functioning normally.

Surveyor 7, which was launched on January 7, 1968, soft-landed on the moon on January 9, 1968, in spite of the fact that it had been given a 40 per cent chance of landing safely on the lunar surface. It was one of the best equipped lunar lander to soft-land on the moon and was *the first to be equipped with a dual method of examining the soil.* It was originally planned that Surveyor 7 would land in the crater Hipparchus about one thousand miles north of Tycho, believed to be one of the most promising landing sites on the moon. However, its course was shifted toward Tycho, because it was considered that this lunar region offered better possibilities as a landing site for the planned Apollo missions. Although the target selected for Surveyor 7 in the Tycho crater was 12 miles in diameter, the spacecraft succeeded in landing close to it and 45 minutes later its television cameras began transmitting pictures. Its primary mission was to determine the nature of the soil below the upper crust by examining the debris on the slope of the Tycho crater, conical in shape with a depth of 15,000 feet and a rim-to-rim breadth of 53 miles.

The Surveyor's mechanical device dug a trench 18 inches in depth and by bombarding it with atomic particles by means of an analyzer box, examined the chemical composition of the soil. At first the analyzer box would not drop into the trench after being released by a light explosive charge but remained suspended by a cord above the trench. Radio command signals were sent to prod and nudge it with the digging arm but it could not be moved. Finally, the box was pressed downward against the side of the Surveyor by the digging arm and dropped into the trench.

This particular mission was basically scientific and provided chemical analyses of the moon's soil and 21,000 pictures revealing striking mosaics near the Tycho crater. It brought to a conclusion the Surveyor program which had soft-landed successfully five spacecraft on the surface of the moon. Together, they had transmitted to earth 88,000 photographs of the various regions of the lunar surface and telemetered the chemical composition of its soil.

The 148-pound Pioneer 9 was lifted off its pad by a three-stage Delta rocket on November 8, 1968, having as its destination a solar orbit. Six hours after take-off, it assumed an orbit with an aphelion of 93 million miles and a perihelion of 70 million miles requiring 297.5 days to complete, with all of its systems functioning satisfactorily. It was accompanied on part of its journey by a 40-pound space capsule which was ejected into an earth orbit to serve as a testing target for the Apollo global-tracking network. This mission was scheduled to carry out eight experiments to record and transmit data relating to the solar wind, cosmic rays, cosmic dust and electrical and magnetic fields. Pioneer 9, together with Pioneer 7 and Pioneer 8, already circling the sun on widely separated paths, were expected to accumulate data to provide the most comprehensive view of the conditions existing in interplanetary space. This information was to be used for warning of solar storms, whose radiation might endanger future manned flights to the moon.

In 1969, two unmanned Mariner spacecraft were launched toward Mars on trajectories if followed would cause them to fly by that planet during mid-summer. Their purposes were to collect information about the Martian atmosphere and other scientific data, and to take television photographs of the planet's surface. Mariner 6 was scheduled for a fly-by within 2,000 miles of the Equatorial Region of the planet and Mariner 7 within the same distance of its South Polar Region. Together their cameras would photograph about twenty per cent of the surface of Mars.

Mariner 6, an 850-pound satellite carrying two television cameras and remote-sensing instruments, was propelled into interplanetary space by an Atlas-Centaur rocket on February 24, 1969. The spacecraft separated from its booster rocket within 15 minutes of launching and headed for a Martian orbit traveling about 25,700 miles an hour. Its performance was flawless and it reached the vicinity of Mars after a 241-million mile voyage through space on July 29, 1969 and began scanning the Equatorial Region of the planet.

The radio command signals to turn on its instruments, were sent from the Jet Propulsion Laboratory at Pasadena, California on the evening of July 29. This took five and a half minutes to reach the spacecraft and another five and a half minutes for the response to reach earth. The first series of 33 pictures was taken when Mariner 6 was 771,500 miles from Mars and continued to 728,000 miles. The two cameras worked alternately every 42½ seconds during the 17-minute drop from 4,800 miles to fly-by at 2,150 miles. The more distant pictures showed the surface of Mars to be covered with craters, extensive mountainous areas, and sharp cliffs. The close-up pictures revealed that the dark equatorial land was not covered with vegetation.

The 910-pound Mariner 7 was sent to Mars by an Atlas-Centaur rocket on March 27, 1969, and reached the vicinity of that planet about a week after Mariner 6. Three temporary malfunctions developed during the early stages of its flight: sudden drop in power, wrong operating conditions of the on-board computer, and drop in radio transmitting power. All were overcome and thereafter the spacecraft operated in a near-flawless manner. The two television cameras took some of the most spectacular photographs of the surface of the South Polar Region and its sensor measured the

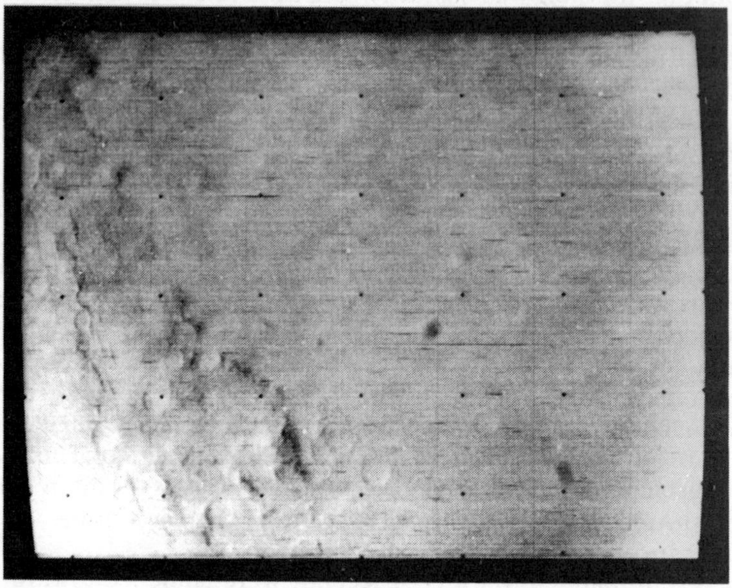

A Wide-Angle View of the Hellespontus and Hellas Regions of Mars Taken by Mariner 7. The structure at the left appears to be a series of ridges and escarpments separating the dark Hellespontus and the area to the west (right) from Hellas. (Source: NASA - Photo No. 69-H-1408)

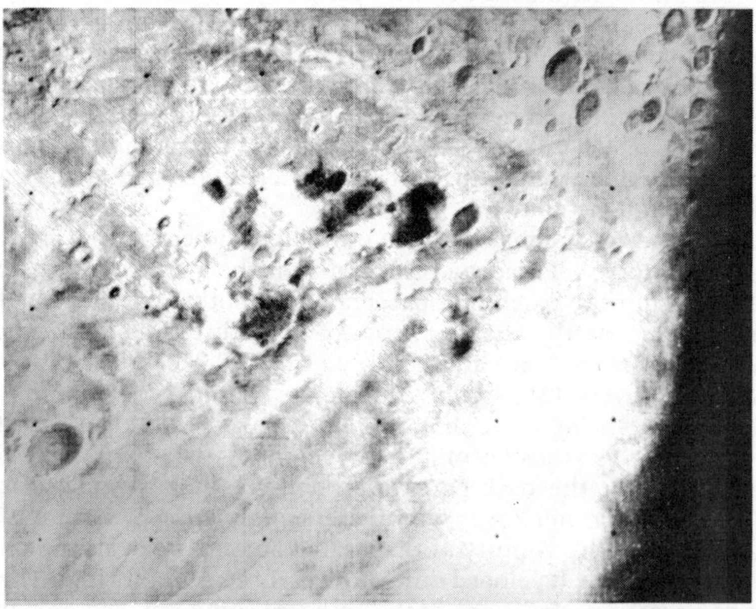

The South Pole Cap Region of Mars Taken by Mariner 7. The photograph shows a wide variety of crater sizes and forms, as well as linear and blotchy features obviously not related to cratering. (Source: NASA - Photo No. 69-H-1382)

variations in the planet's surface. From within 2,200 miles on August 5, 1969, its sensoring device indicated that the Martian polar snow might be formed of frozen carbon dioxide.

The Mariner Project for 1971 was designed to send two identical 2,200-pound spacecraft to Mars to orbit the planet and map about seventy per cent of its surface by a photographic surveillance in search for possible landing sites for future unmanned or manned spacecraft. The Mariner spacecraft was a sun-powered electronic laboratory equipped with four solar panels containing 4,368 solar cells for converting sunlight into 500 watts of electricity. It had three antennas for transmitting and receiving radio signals. Computers, combining programmed information, controlled the timing and sequence of the spacecraft's systems. It was equipped also with two cameras, one for wide views and the other for narrow or close-up views along with other sensitive instruments.

The first spacecraft of this dual mission was to swing into a Martian orbit from 750 miles to 10,000 miles above Mars at an angle of 80 degrees to its equator and circle the planet every 12 hours with its cameras focused on broad areas of the surface. The second was expected to enter a Martian orbit, with altitudes ranging between 530 miles and 20,500 miles above the planet, at a 50-degree angle to equator of Mars and circle the planet every 20½ hours. Its cameras were to be focused on smaller areas of the Martian surface of particular interest to the space scientists. Together the two were to chart the planet's strange phenomena including the yellowish dust storms, blue haze, and frosty polar caps, and search for clues as to its capabilities for supporting some form of life.

Mariner 8 lifted from its pad in a flawless launch on May 8, 1971 but failed to attain a Martian orbit. Shortly after take-off, it tumbled out of control because of a malfunction in the second stage of the carrier rocket's guidance system. It fell into the Atlantic Ocean about 900 miles down range from the launching site or about 350 miles north of Puerto Rico.

The launching of Mariner 9 was postponed in order to avoid repetition of the events surrounding the aborted flight of Mariner 8. On May 30, 1971, it lifted off on the 247-million mile journey to the vicinity of Mars. *On November 13, 1971, it went into orbit about Mars becoming the first man-made object to circle that planet and the first to take close-up television pictures of its surface.* The radio signals confirming the successful orbiting maneuvers took 6 minutes and 43 seconds to reach the ground-control station. Mariner 9 first swung into orbit over the planet's southern hemisphere at an altitude of 940 miles above the Sirenum Sinus, a Martian feature appearing dark when viewed from earth. The spacecraft's planned orbit was to take the vehicle within 868 miles of the surface of Mars and as far away as 11,135 miles as it moved between Phobos and Deimos, Martian moons, requiring 12 hours and 34 minutes to complete. When it passed behind Mars, radio contact was lost for 32 minutes and when it emerged it was only thirty miles off the predicted altitude. If Mariner 9 continues in its orbit, it will not fall to the surface of Mars for at least 17 years.

The onboard tape recorder began playing back pictures taken within a range of 130,000 to 700,000 miles, but a great dust storm prevented Mariner 9 from taking clear and well-defined pictures, as it approached the planet, from November 13 until January 12, 1972. The cameras then began to transmit high quality pictures revealing considerable detail of the planet's deeply pitted surface, with its extensive leopard spot patterns, and canal-like canyons. The spots appeared in sizes ranging from 100 miles or more in width down to the smallest size observable by the cameras. On

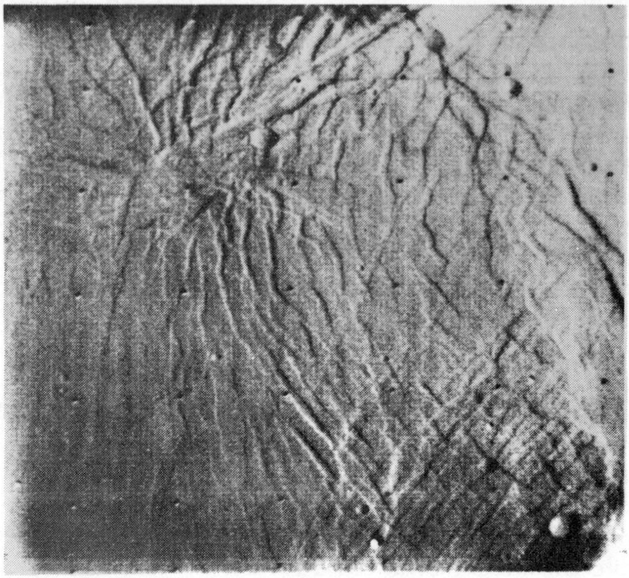

The Phoenicis Lacus Area of Mars Taken by Mariner 9. The fault valleys are about 1½ miles across and the plateau lies about 3½ miles above the mean elevation of Mars. The surface is relatively young and may have been covered by volcanic deposits and later broken into faults that cut the rocks into mosaic-like fragments. (Source: NASA - Photo No. 71-H-1929)

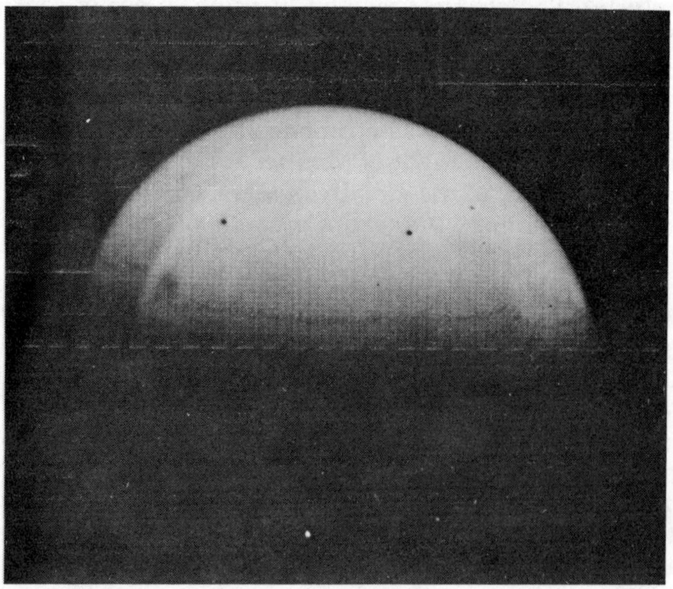

Nix Olympica, a Curious Ring-Shaped Feature of Mars Taken by Mariner 9. The darkish spot near the top of the photograph has been identified with the feature photographed by Mariner 6 and 7 in 1969. This is the most conspicuous feature observed on Mars by Mariner 9 and is a point which radar indicated as one of the highest on the planet. It may be a high mountain or plateau viewed as it rises through the bright dust which surrounds the rest of the planet. (Source: NASA - Photo No. JPL P-12672)

January 12, 1972 a panoramic photograph covering an area of about 235 by 300 miles and taken from a distance of 1,225 miles, was transmitted showing a vast chasm 75 miles wide and 300 miles long, with branching canyons, resembling a network of dry river beds.[11] From the time Mariner 9 went into orbit until mid-January, it transmitted 4,000 pictures, some of which were blurred by the great dust storm. Mariner 9 completed the major part of its assignments and continued to orbit Mars as the planet moved farther away from the earth. This mission may prove to be one of the most fruitful scientific missions conducted in connection with Mars.

The Viking Program is one of the most ambitious unmanned space exploratory projects undertaken by the United States. Its primary purpose is to search for evidence of life on Mars and to test the Martian atmosphere for the presence of argon and other gases.

Two identical Viking spacecraft were sent to Mars during 1975. The two missions have similar purposes but will explore different regions of Mars. The first mission will search with heat- and water-sensors for the best landing site for the spacecraft's lander on the plains of Chryse located near a channel. This channel may have been carved by flowing water hundreds of millions of years ago. The second mission will search Mare Acidalium in a similar manner for the best landing site in the Cydonia section near the southern limits of the planet's northern polar cap.

The sophisticated Viking spacecraft is a highly instrumented combination orbiter and lander. The orbiter has been nicknamed the *Pathfinder* and the lander the *Prospector.* The former weighs 5,100 pounds and the latter 2,400 pounds. The orbiter will orbit Mars after ejecting its lander and serve as a mothership to it after the lander has landed on the surface of Mars. It will receive signals from the lander and relay them to earth. The lander is a crawling space laboratory designed to search for microscopic and plant life by analyzing the soil samples scooped by its ten-foot boom forming the third leg of its tripod, and placed in the miniature laboratory where their organic composition will be analyzed for metabolism. The results will be transmitted to the orbiter and in turn relayed to the Viking Mission Control at the Jet Propulsion Laboratory, Pasadena, California. The signals will take about twenty minutes to reach the earth and the exploratory period may last for a period of ninety days or more.

The first of the two Viking spacecraft was scheduled to lift off the pad at Cape Canaveral on Monday, August 11, 1975. Because of a sticky valve in the Titan 3E Centaur carrier rocket, the lift-off was delayed until Thursday August 14. It was again delayed because a switch had shifted from *off* to *on* depleting two of the spacecraft's batteries. This occurred while the faulty valve was being replaced, preventing the engineers from monitoring the spacecraft.

The Viking (A) spacecraft was replaced by the Viking (B) spacecraft on top of the carrier rocket. The Viking (B) spacecraft was to be used in the second mission which had been postponed until September 1. The Viking (A) spacecraft was returned to the Kennedy Space Center to be "cleaned" and prepared for the second mission. This exchange reduced the time in getting the first mission under way.

Efforts were made to shorten the preparatory period in hopes that the first mission could be initiated on August 20. Furthermore, time was important because the sun would pass between Mars and the earth in November 1976, interrupting communication for several weeks. It was essential to have the Viking spacecraft in the vicinity of Mars before that time. If the period of preparation can be shortened, there is still the chance of the lander landing on Mars on July 4, 1976, as originally planned when the lift-off was scheduled for August 11.

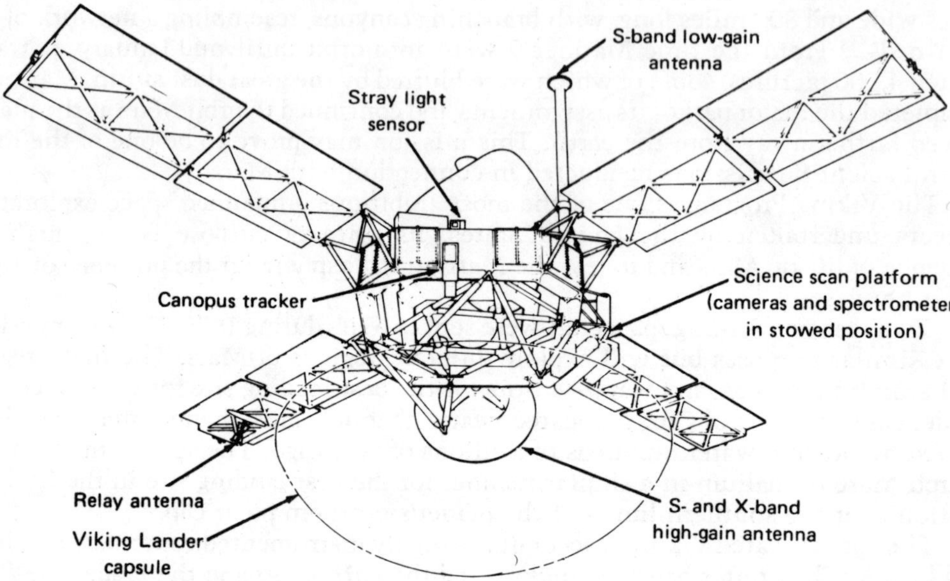

Bottom View of the Viking Orbiter Launched August 20, 1975 (Source: NASA - Photo No. 75-H-446)

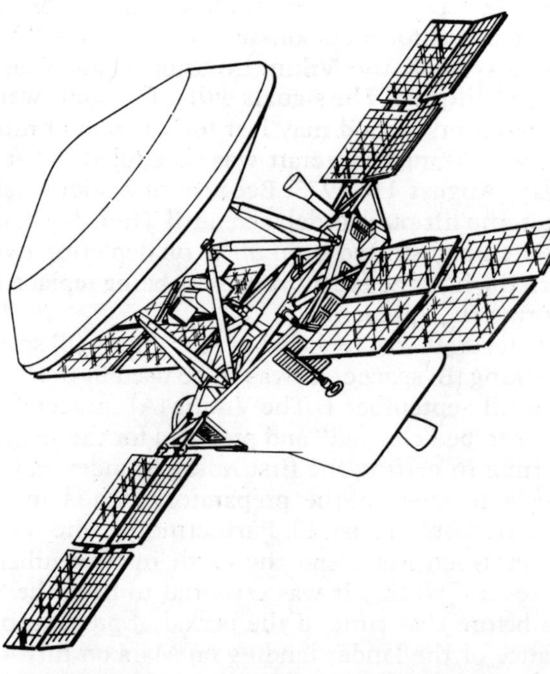

Artist Concept of the Viking Orbiter and Landing Capsule (Source: NASA - Photo No. 75-H-458)

Finally on Wednesday, August 20, at 5:22 p.m. (EDT) the Viking-I spacecraft was blasted into space by the Titan 3E Centaur carrier rocket, the most powerful launch vehicle possessed by the United States since the retirement of the Saturn V. The initial earth-parking orbit was 105 miles high and about thirty minutes after lift-off, the Centaur stage rocket fired, injecting the spacecraft into a Mars trajectory. Although the greatest distance between Mars and earth in a direct line is nearly 230 million miles the trajectory taken by the spacecraft in looping around the sun is 505 million miles and will take 303 days to complete.

Viking-I is scheduled to orbit Mars on June 19, 1976, survey sites on Mars, and land its lander at a site called Chryse on July 7, 1976. Originally it was assumed that 18 days would be needed to trim the Martian orbit and survey landing sites. If this period can be reduced it may still be possible to land the lander on the planet on July 4, in commemoration of the Nation's Bicentennial Year.

The launching of Viking II, previously postponed because of the delay in launching Viking I, took place at Cape Canaveral on Tuesday, September 9, 1975 at 2:39 p.m. (EDT) and in less than an hour it was injected into a Martian trajectory. On Wednesday, it was estimated to be 4.8 million miles behind Viking I, launched on August 20, 1975. Viking II is scheduled to land its lander on Mars one year from the date of its launching.

ADDENDUM

Viking I began orbiting Mars on June 19, 1976, and there was hope that its lander would touchdown on Mars on July 4 for a "Bicentennial landing." Surveys of the proposed landing site revealed that it was too rough and search was started for an alternate landing site. Finally, on July 20 the seventh anniversary of the first landing of men on the moon, the Viking I lander made a sensational touchdown on Mars and immediately began transmitting clear and distinct pictures of the physical features of the planet's surface.

REFERENCES

1. John Wilkins, *The Discovery of a World in the Moon,* Proposition VIX, pp. 132-133.
2. This chapter is based primarily on information obtained from the *Semiannual Reports of the National Aeronautics and Space Administration to Congress,* leading newspapers, including *The New York Times* and *Aviation Week and Space Technology.*
3. Carsbie C. Adams, *Space Flight, Satellites, Spaceships, Space Stations and Space Travel Explained,* (New York: McGraw-Hill Company, Inc., 1958), pp. 136-7.
4. The Pioneer Project was initiated to study the major characteristics of space including radiation, magnetism, micrometeorite impacts, and other scientific data. There have been 11 launchings under this project, the most recent on April 6, 1973, (pp. 68-70).
5. The Ranger Project was initiated to investigate the surface of the moon by unmanned spacecraft as they approached it and by landing survivable payloads on its surface. There were nine launchings under this project, the last on March

21, 1965 (p. 48). These missions supplied more than 17,000 lunar pictures, most of them up to 2,000 times clearer than those obtained by earth-based telescopes.

6. The Mariner Project was initiated to explore the planets of Venus and Mars and their near-spatial environments. There have been ten launchings, the most recent on November 3, 1973 (pp. 70-76).
7. Pioneer 6 was the first of a series of four Pioneer spacecraft launchings scheduled to be launched at six-month intervals to explore interplanetary space to chart future courses for astronauts who will be traveling between planets.
8. The Surveyor Project was more sophisticated than the Ranger Project. It was designed also for the exploration of the moon. The unmanned spacecraft were more complicated and designed to soft-land on the moon's surface and to conduct on-the-spot scientific surveys of its physical features. There were seven launchings under this project, the last on January 7, 1968 (pp. 54-55).
9. The National Aeronautics and Space Administration approved the Lunar Orbiter Project on August 30, 1963. The project differed from the Surveyor Project in that it relied on spacecraft in lunar orbit for assembling and transmitting data relating to the moon and its environment. It was the Nation's first effort of photographic reconnaissance and mapping of the lunar surface and monitoring radiation and other phenomena in its vicinity. Five lunar missions were launched, the last on August 1, 1967 (p. 53).
10. Pioneer 6, Explorer 33, three orbiting Geophysical Observatories and Mariner 4. These spacecraft should all be in position by mid-October, 1969, to monitor simultaneously events in a strip of the solar surface extending halfway around the sun.
11. The information gleaned from the pictures indicates that two geological processes have shaped the Martian surface: volcanic eruptions and collapsing of its surface.

Chapter VI

UNMANNED PLANET-PROBING FLIGHTS OF THE UNITED STATES TO JUPITER-SATURN AND VENUS-MERCURY[1]

Plans were being made to explore other planets in the solar system while the explorations of Venus and Mars were in progress with special consideration being given to Jupiter, the fifth and largest planet of the solar system. This huge planet, the gateway to other planets, has a diameter ten times that of the earth and a mass 318 times as great. It holds the interest of scientists because it is believed to consist of a huge mass of highly activated gases and liquids. The planet is obscured by a cloud cover above which is a Great Red Spot, measuring 30,000 miles long and 8,000 miles wide. It also gives off more energy than it absorbs from the sun but the source of the excess energy remains unknown.

Pioneer 10, a 570-pound spacecraft was scheduled originally to lift off on its extended journey to Jupiter on February 27, 1972. Less than an hour prior to the time of its launching, the flight was postponed, until February 28, because of high winds and a temporary failure of power at the launch pad. It was delayed again, this time for 48 hours because it was not possible to reprogram the guidance computer of the booster rocket to compensate for the high-altitude winds. The lift-off was rescheduled for March 1, at which time its departure was delayed again for 24 hours because of high-altitude winds. Finally on March 2, 1972, after a delay of 25 minutes, Pioneer 10 was hurled into space on its way to Jupiter. At the time of launching, it was anticipated that it would reach the vicinity of Jupiter by Christmas 1973.

The primary objectives of the mission were to: survey the density of the asteroid belt between Mars and Jupiter, take close-up pictures of the planet, transmit to earth data relating to Jupiter's magnetic field, radiation belts, temperatures and composition of its atmosphere, investigate cosmic rays and search beyond Jupiter for the boundary where the force of the solar wind stops and interstellar space begins.

It was hoped that Pioneer 10's radio transmitter and receiver would continue to function until at least 1977 when the spacecraft should cross the orbit of Saturn, a billion miles from the sun. Thereafter it would escape the solar system and radio contact with it would be lost because of the vast distance from the earth. By 1980, Pioneer 10 may have entered the galaxy of the Milky Way and assumed theoretically an endless straight-line path in interstellar space.

The Pioneer Spacecraft Used for Exploring Mars, Jupiter, and Saturn

The power of the Atlas-Centaur carrier rocket was increased to enable it to impart a velocity of 31,122 miles an hour to the spacecraft. This velocity was necessary for it to escape the earth's gravitational force and to be injected into a Jupiter trajectory. As Pioneer 10 approaches the planet, its speed should accelerate from 20,000 to 78,000 miles an hour because of the gravitational pull of Jupiter. This would be a speed no other man-made satellite has yet attained. After the spacecraft passes Jupiter, the high velocity will fling it into a path toward Saturn and beyond.

Pioneer 10 was equipped with about sixty-five pounds of instruments consisting of sensing devices capable of mapping and measuring the planet's magnetic field, determining the size and velocity of asteroid particles, and investigating the high energy radiation belts, radio emissions and atmospheric gases. It was not equipped, however, with life-detecting devices, but the nature and extent of the data transmitted by Pioneer 10 may provide some clues to this perplexing question.

The spacecraft was designed to spin as it traveled forward providing a circular field every five minutes for its sensing devices to scan. The electrical power for energizing the onboard systems was to be derived from the decay of radio-active plutonium.[2] It was necessary to utilize this method of power generation because of the vast distance the spacecraft would be from the sun when it reached the vicinity of Jupiter and went beyond, making it impractical to rely on solar energy. It was also equipped with a nine-foot dish antenna enabling it to transmit and receive radio signals over the great distance the spacecraft would cover on its journey in space.

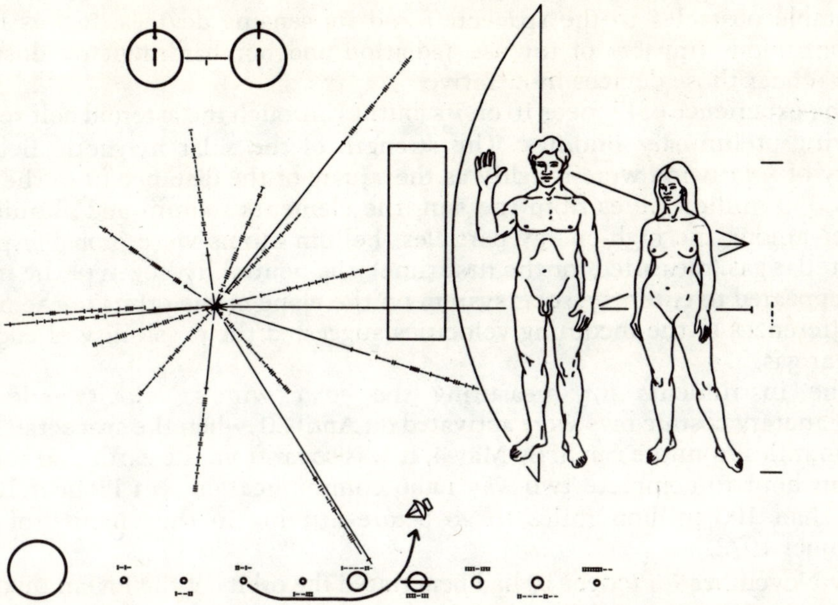

The Plaque Attached to Pioneer 10, Showing Symbolic Message for Educated Inhabitants of Some Other Star System Who Might Intercept it Millions of Years from Now. The design is etched into a gold-anodized aluminum plate 6" x 9" attached to the spacecraft's antenna support strut. (Source: NASA - Photo No. 72-H-192)

When it reached the vicinity of Jupiter, radio signals traveling at the speed of light would take 45 minutes to reach the earth. In other words, 90 minutes would be required to transmit a radio signal from earth to Pioneer 10 and receive a response from it.

The spacecraft had attached to it a six by nine inch aluminum plaque annodized with gold upon which was inscribed a message depicting, in scientific symbols, who sent the spacecraft and where they lived in the solar system. *This was the first direct attempt by man to communicate with other intelligent beings, if they should exist, somewhere in outer space.* The symbols were based on the energy difference between the two basic states of the hydrogen atom (the universal element) as units of time and distance. It was hoped that they would have meaning for those who should chance to recover the plaque and enable them to decipher it.

The plaque showed the position of the earth in terms of 14 pulsars or celestial sources of radio pulsations, each identified by its characteristic rhythm. It also depicted the inhabitants of the earth—a man and a woman—standing beside the symbolic representation of the spacecraft. Other symbols indicated the position of the earth within the solar system.

The ground control station at Mountain View, California, was out of radio contact with Pioneer 10 on January 15 through January 17, 1973 because at that time the sun was between it and the earth.

On February 15, 1973, Pioneer 10 emerged undamaged from its seven-month passage through the asteroid belt. *This was the first time that a man-made satellite had traversed the asteroid belt between Mars and Jupiter.* This accomplishment set at rest the concern of some space scientists that the belt would present insurmountable obstacles to the spacecraft and its sensing devices. It was feared that the continuous impacts of intense radiation and bombardment by dust particles would render those devices inoperative.

The experience of Pioneer 10 on its journey through the asteroid belt revealed the following preliminary findings: The strength of the solar magnetic field and the density of solar wind were roughly as the square of the distance from the sun as far out as 350 million miles from the sun; the elements sodium and aluminum were present among the high-energy particles; helium atoms were probably part of the interstellar gas (measured for the first time); the neutral hydrogen of the interstellar wind appeared to enter the solar system on the plane of the orbits of the planet; and the differences of the incoming velocities suggested the possibility of eddies in the instellar gas.

The instruments for measuring the solar wind's velocity and sampling interplanetary cosmic rays were activated on April 20, when the spacecraft was seven million miles from the earth. By May 4, it was so far from the earth that it took more than an hour to complete two-way radio communication, yet Pioneer 10 still had more than 100 million miles to go before arriving in the vicinity of Jupiter in December 1973.

By November 9, Pioneer 10 had penetrated the orbits of the Jovian moons, where it began to encounter a region of intense dust particles surrounding the planet, as well as the Jovian radiation belt believed to be the most intense in the solar system. The preliminary findings, based on the data received from the spacecraft, indicated that the dust intensity of this belt was 100 times more than expected. Nevertheless, the sensing instruments survived and continued to transmit information and on November 25, began sending back images of the Great Red Spot. During the sequence observations beginning on November 3, the planet was scanned eight hours daily but from November 26, to December 10, the image transmission was continuous.

On November 26, the spacecraft was reported to have penetrated the Jovian sphere of influence, extending some 4,790,000 miles outward from the planet's cloud top and it was given the final rocket thrust preparatory for the fly-by of Jupiter on December 3, 1973. The crossing of the bow shock wave on November 27, was evidence that it had passed into its sphere of influence.[3] At that point, there occurred an abrupt decrease in the velocity of the solar wind and a sharp increase in its temperature. The bow shock wave was encountered about twenty-four hours sooner than anticipated because of the planet's strong magnetic influence.

Pioneer 10 was 4,200,000 miles from Jupiter when on November 27 it crossed into the planet's true magnetic field, called the magnetosphere, where the solar wind does not penetrate. The estimated diameter of this magnetic field was in excess of 8,000,000 miles in contrast to that of the earth equal to about 80,000 miles. Jupiter is the only planet (other than the earth) in the solar system, known to possess an intensive magnetic field and radiation belts of particles trapped and accelerated by a magnetic field. The spacecraft was expected to penetrate the most intensive of Jupiter's radiation belts about six hours before its closest approach to the planet's surface.

It was estimated that the spacecraft was 2.4 million miles from Jupiter on November 30, traveling at 26,200 miles an hour. As it approached closer to the planet, the transmitted images grew larger and the multicolored bands and the Great Red Spot became more distinct. *Pioneer 10 was the first spacecraft to explore Jupiter, the largest planet of the solar system.*

Jupiter's Great Red Spot and a Shadow of Its Moon, Io, Together with Jupiter's Cloud Structure, as Viewed from 1,580,000 Miles of the Giant Planet by Pioneer 10. (Source: NASA - Photo No. 73-H-1164)

The findings up to that time produced two surprises for space scientists: the distance to which the trapped radiation particles extended from Jupiter, estimated at 4.2 million miles, and the fact that there was no rapid build-up in the intensity and density of the radiation as the spacecraft approached nearer to the planet. These observations caused considerable puzzlement on the part of many space scientists. Furthermore, the first estimate of the strength of the Jovian magnetic field was forty times that of the earth but that estimate was reduced later to eight times.

As the spacecraft sped toward its rendezvous with Jupiter, it crossed the orbit of Ganymede, Jupiter's largest moon, coming within 277,000 miles of it. Its infrared radiometer returned data indicating that its temperature was 235 degrees below zero (Fahrenheit). It also passed close to Europa, a small satellite of Jupiter. On December 1, it was 1.6 million miles from the planet, traveling at about 29,000 miles an hour and on December 2, it was 900,000 miles from it.

The ground control station experienced some concern about six hours before the spacecraft reached its nearest point to Jupiter. The onboard imaging system suddenly ceased transmitting signals and changed to another mode of operation. The controllers switched it back to its normal operation, but not before losing data, including an image of Io, a Jovian moon.[4]

On Monday, December 3, 1973, Pioneer 10 passed within 81,000 miles of Jupiter traveling at 82,000 miles an hour (23 miles a second). It had *traveled farther and faster than any man-made satellite up to that time*. By December 5, 1973, it was 1.7 million miles beyond Jupiter and its speed had dropped to 29,000 miles an hour. It was on a course leading past Saturn into outer space, beyond Pluto and probably, by 1987, on a galactic path in the direction of the constellation Taurus. If Pioneer 10 continues on this course, it would be the first spacecraft to escape from the solar system.

It was discovered that the average temperature of Jupiter's cloud cover was 210 degrees below zero (Fahrenheit). A release on April 9, 1974, of the results of a four-month study of the photographs returned by Pioneer 10 indicated that the Great Red Spot was a towering mass of clouds. The pictures also showed a smaller spot similar in color and shape.

A further interpretation of the data and photographs transmitted to earth by Pioneer 10 and released months later,[5] indicated that: Jupiter had an equatorial diameter of 88,298 miles which was 6,000 miles greater than the spread between its poles; it was a huge mass of highly activated gases and liquids with no solid surface; the atmosphere of its outermost 600 miles was composed of hydrogen and helium gases, interspersed with clouds consisting of ammonia, ammonium hydrosulfide and ice; the rest of the planet was made up of liquid hydrogen, except for a possible iron-bearing core, having a temperature of about 54,000 degrees (Fahrenheit); the hot grey-white rising clouds circling the planet were formed into bands because of their rapid rotation; and the dark orange-brown bands running parallel to the lighter bands were probably caused by descending gases. It is questionable whether such a hostile environment, interspersed with intense bolts of lightning and disturbed by winds with velocities in excess of 300 miles an hour, would be conducive to life as we know it.

This unusual spacecraft survived the passage through the intense radiation belt of Jupiter. As of early December, 1974, two years and nine months after being launched from earth, Pioneer 10 was still transmitting scientific information back to earth.

Pioneer 11 was launched on April 5, 1973, as a back-up mission to Pioneer 10 with the expectation that it would reach the vicinity of Jupiter sometime in December 1974. The spacecraft, launched with an escape velocity of 31,100 miles an hour, assumed a near-perfect Jupiter trajectory and within 11 hours after lift-off, flew past the moon on a coasting trajectory to Jupiter. It was to approach within 27,000 miles of the planet in contrast to 81,000 miles achieved by Pioneer 10.

The spacecraft, although identical to that used in the Pioneer 10 mission, was equipped with an additional magnetometer to measure more adequately the Jovian magnetic field. There were 14 experiments designed to relay data to earth relating to interplanetary space, interstellar gases, cosmic rays, and meteoroids.

There was concern in connection with the failure of one of the two booms holding the nuclear power unit, to deploy to its full length of nine feet, as the spacecraft traveled in space. Although not viewed as a serious problem, it was resolved on April 6, following an overnight analysis.

On April 8 when the spacecraft was 1.4 million miles from the earth, ground control activated the onboard cosmic telescope designed to monitor cosmic ray particles from the sun and interplanetary space. It will be used to monitor high-energy particles and the radiation belt when Pioneer 11 arrives in the vicinity of Jupiter.

On April 11, the course of the spacecraft was altered to keep open alternatives for the fly-by of Jupiter. After passing the planet it may be directed to follow a course leading to outer space, enter a solar orbit, or follow a path leading to a fly-by of Saturn. It could also be placed in a trajectory for a fly-by of the moons of Jupiter.

The spacecraft's radio transmission was switched to a back-up amplifier on May 18, because of the erratic changes in the output power of its main systems tubes. On

July 7, Pioneer 11 crossed the orbit of Mars, marking the second time that a spacecraft had accomplished that feat. It was then 54 million miles from the earth, traveling at 65,000 miles an hour, and by December 8, was traversing the asteroid belt beyond Mars.

It was decided on March 24, 1974, to change the course of Pioneer 11 to permit it to fly-by Saturn, after its fly-by of Jupiter. This change should bring it within 26,600 miles of Jupiter by early December, 1974, and allow the spacecraft to approach the planet from below its south pole and pass behind it after which the gravitational force of Jupiter would reduce the spacecraft's speed and inject it into a long looping solar orbit.

After a journey through space of 21 months, traveling at 81,000 miles an hour, Pioneer 11, on December 1, 1974, was nearing the climax of its 620-million mile voyage culminating in its fly-by of Jupiter. At that time, it was one million miles from the planet. It was only hours away from its encounter with the intense Jovian radiation belt and its imaging system continued sending high-quality pictures of the planet's polar region. It transmitted pictures of Jupiter and crossed the orbit of Callisto, one of the four large moons of Jupiter, considered a most likely landing site for a possible manned mission to the vicinity of Jupiter, because of its low level of radiation.

On December 2, Pioneer 11 crossed the orbit of the Jovian moon, Ganymede, and moved on to cross the orbit of Europa, and entered the planet's radiation belt. During the evening, it flew past Io and a lesser moon, Amalthea, and on toward its rendezvous with Jupiter.

At 9:22 p.m., (PST) on December 3, 1974, the spacecraft dropped to within 26,600 miles of the surface of Jupiter, its nearest approach to that apparently boiling and gaseous planet. Signals were received at 10:24 p.m. indicating that it had survived the planet's intense radiation belt and was on its way toward Saturn.

The signals came a little more than an hour after contact was lost with the spacecraft as it went behind Jupiter attaining a speed of 107,000 miles an hour relative to the planet, *the fastest speed that a man-made satellite had ever flown*. For 42 minutes, including the time it was nearest the planet's cloud top, it was behind Jupiter. In addition, the complex electronics of the imaging device made it necessary to shut it off for a half-hour because of the intensity of the radiation belt. *No spacecraft had ever been exposed to such intensity*, but because of the short duration of its exposure, it suffered only minor damage to its systems. The scientists had expected the spacecraft to be struck at least five times by micrometeorites, but fortunately it was hit only twice.

Following a corkscrew-like path, Pioneer 11 left the northern region of the giant planet four hours later. During this period, it transmitted the best pictures of the Great Red Spot and for the first time sent back pictures of Jupiter's dark polar region.

Pioneer-Saturn continued on toward Saturn, a yellowish hued planet estimated to be seven hundred times the size of the earth and surrounded by three concentric rings, believed to be composed of flying debris. The space between the innermost ring and the planet itself is believed to be 16,000 miles, but this space should be cleared of the debris by the gravitational force of Saturn.

The recurrence of orange in the pictures received from Pioneer 11 is believed by some to be indicative of some form of living organism on the planet, whereas others rejected that conclusion. One possible explanation for the recurrence of orange is that there were present in the atmosphere organic molecules combined into living organisms. If true it would appear to conform with the theory of the chemical evolution of life and would be supported by the fact that its atmosphere contains ammonia, methane and hydrogen, together with water. These are the elements believed to have produced life on earth some four billion years ago. The other explanation is that there were clouds of ammonium hydrosulfides, which were orange in color.

It was decided that the gravity of Saturn would not be used to send Pioneer 11 on an extended journey toward Uranus, but to put it on a course bringing it closer to the equatorial plane of Saturn. This would enable it to fly by Titan the largest moon of Saturn, with an orbit of about 76,000 miles from it. It is believed that the atmosphere of Titan might be capable of sustaining some form of life. To assure that its native environment will be preserved until it can be further explored, there is an international agreement to the effect that precaution will be taken to avoid approaching too closely for fear of contaminating it with earth microorganisms. Pioneer 11 was therefore scheduled to fly-by within some 12,000 miles of Titan. The basic reason for not attempting to fly-by Uranus was that the spacecraft's nuclear-powered electric plant would be so weak by the time it reached that planet that it would not possess the capability of operating all the spacecraft's systems simultaneously.

By September, 1979, Pioneer-Saturn, should rendezvous with Saturn and later with Titan about six and a half years after being launched from earth. The latter phase of the flight is speculative in that the spacecraft would be nearing the end of its productive life and might be subjected to destructive bombardment by large particles of debris within the concentric rings of Saturn.

Interest was also directed to the exploration of Mercury, the smallest planet in the solar system and nearest to the sun, with a diameter of about three thousand miles, or about a third of the earth's diameter. Its surface is about one-seventh of that of the earth; its volume six-tenths, and its mass one twenty-fourth. The average distance of Mercury from the sun is 36,000,000 miles with a range from 28,000,000 to 43,000,000 miles.

Mariner 10, a 1,110-pound unmanned spacecraft, was hurled into space on a flight to Mercury on November 3, 1973. *This was the first spacecraft to be sent on a reconnaissance mission to Mercury*. It was estimated that the journey would take five months and that the spacecraft would fly by Venus on its way to Mercury. *This was also the first attempt by the United States to achieve, in a single space mission, the exploration of more than one planet, by using the gravity of another planet as a propelling force.*

Mariner 10 had two solar panels, each nine feet long and three feet wide, to provide power for the instruments and the spacecraft's systems. When going toward the sun, it required only two of the four solar panels for power, but when receding it required all four. It was also equipped with thermal blankets, a deployable sunshade, thermostatically controlled louvres, and was coated with white paint to reflect excessive heat. These precautions were necessary because the spacecraft was to operate within 36 million miles of the sun and be subjected to intense polar radiation. In addition, it was equipped with two telescopic cameras, one fitted with wide angle lenses and the other with lenses for high-resolution photography. They were mounted on rotating scanning platforms to provide greater surveillance.

The spacecraft depended on the gravitational force of Venus in reaching Mercury. This force caused a bend in the trajectory and reduced the speed of the spacecraft, allowing it to intersect with the orbit of Mercury. The use of the gravitational force of

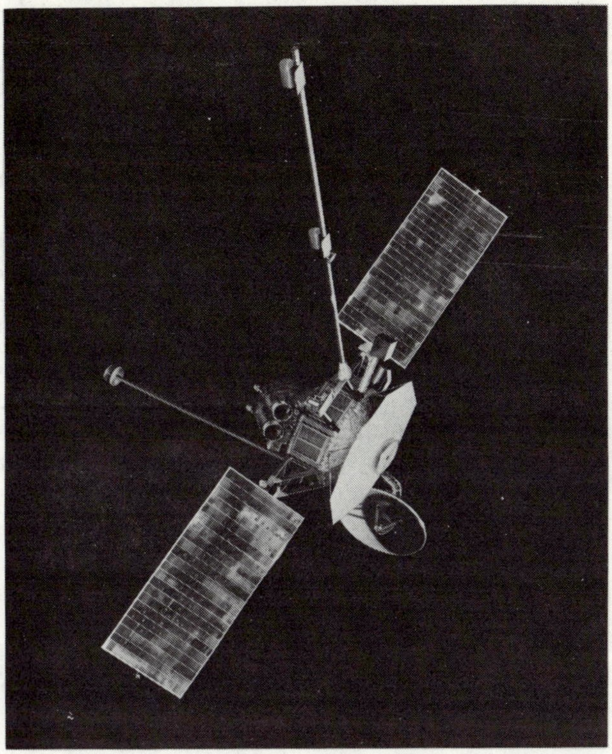

In-Flight Configuration of Mariner 10 Used for Exploring Venus and Mercury (Source: NASA - Photo No. 73-HC-816)

one planet to reach another, without additional rocket power, has opened up the possibility of flying to several planets of the solar system with present-day carrier rockets.

Mariner 10, after coasting in earth orbit for 25 minutes, was injected into a Venus trajectory, and escaped from the earth's gravitational force at a speed of 26,280 miles an hour (7.3 miles per second). It left earth orbit in a direction opposite to the motion of the earth about the sun, causing it to fall toward the sun and gather momentum from the pull of solar gravity. It was anticipated that the spacecraft would approach Venus at 82,440 miles an hour (22.9 miles per second) taking it past the planet at an altitude of 3,300 miles. In passing Venus, it was planned to bend the trajectory by 34 degrees and slow its velocity to 72,000 miles an hour (20 miles per second). This would direct it toward the sun so that its path would intersect with Mercury's solar orbit. Without this bending, Mariner 10 would coast in the direction of the earth.

It was essential that ground control stations monitor closely the trajectory of Mariner 10. An error of one mile in the rendezvous with Venus would result in an error of about 1,000 miles with respect to its planned rendezvous with Mercury. With careful monitoring the spacecraft was to make its closest approach to Venus on February 5 and was to fly by Mercury on March 29, 1974 at an altitude of 621 miles. *This would bring Mariner 10 closer to the sun than any previous spacecraft.* Its close-up observations were scheduled to extend over 19 days, after which the spacecraft would orbit the sun and have a second encounter with Mercury during

September, 1974. *This was the first two-planet mission ever attempted and was the first attempt to use the gravity of one planet as a source of energy to reach another planet.*

The mission had as its primary objectives the collection and transmission of data relating to the shapes, magnetic fields, temperatures and atmospheres of Venus and Mercury and the observation of the characteristics of the Mercurian terrain. In addition, its cameras were to be focused on the Comet Kohoutek on its way to Venus. The major target of the mission was Mercury and it was anticipated that the information obtained and transmitted by the spacecraft's scientific devices would indicate the nature of its surface and alterations, since the formation of the planet billions of years ago.

By November 4, when the spacecraft was nearing 600,000 miles on its journey, Mariner 10's cameras took sharp pictures of the moon. Its two television cameras were cooling because of the lack of heat from the heaters, which had failed shortly after take-off. Fortunately this difficulty was remedied by not turning off the tubes of the cameras thereby generating sufficient heat to compensate for its loss. Of the eight onboard experiments only one failed and that one was least related to the interplanetary study.

In spite of successive technical problems, the spacecraft's systems were functioning satisfactorily as it approached Venus. The power of the transmitter's radio signals was weak but still strong enough to transmit television pictures and data back to earth. The back-up power system took over on January 8 unexpectedly and continued to operate flawlessly.

There was considerable concern on January 29, when one of the gyroscopes began to operate erratically. This malfunction resulted in the loss of about twenty-five per cent of the spacecraft's fuel because the attitude control system had to maintain stability of the spacecraft by compensating for the erratic function of the gyroscope. It was shut off and the spacecraft's navigational system was locked on the sun and the star Canopus. There was also concern that the sensor, fixed on the star Canopus, might receive too much light from Venus as the spacecraft approached nearer to it. Since stabilization of Mariner 10 could no longer be maintained by its gyroscopes, too much light, if leaked into the sensor, might throw the spacecraft off course turning its cameras away from Venus during the critical stage of the mission. Fortunately this did not occur.

On February 5, Mariner 10 swept past Venus approaching to within 3,600 miles of its surface, with its narrow angle cameras covering an area of about twenty miles in width. It sent back pictures showing that the planet was shrouded by a haze below which appeared a brilliant top-deck of clouds. The variations in the intensity of the light suggested some sort of structure, the nature of which may be identified when intensified by computer processing. In addition to the pictures, the infrared sensor provided a profile of the cloud-top covering the surface of the planet facing the sun. The spectrograph scanned the planet revealing the ultra-violet glow of hydrogen in the upper atmosphere and also of helium.

Although most of the several thousand pictures were taken during the fly-by, others were taken as Mariner 10 receded from Venus. They revealed a system of bands and streaks appearing to be somewhat parallel to the Venusian equator. They are believed to have some relationship with the planet's symmetrical spiraling winds thought to be carrying energy from its equatorial to its polar regions. It appeared that the lower atmosphere of Venus was so dense that its surface pressure was 90 times greater than that on earth. This atmospheric density caused the storing up of so much heat that the night side of Venus was as hot as its day side.

An early analysis of the pictures received from Mariner 10, relating to Venus, support the theory that Venus was formed and matured in a manner different from that of the earth. It had been apparently barren and intensely hot from the beginning. The impressions formed from a study of the pictures and data were that the upper atmosphere of Venus was primarily carbon dioxide and constantly fed gas by the solar wind; Venus had virtually no magnetic field to counter the solar wind; the hydrogen, detected in its upper atmosphere, did not, appear to be the product of the breaking down of water evaporating from within the planet but derived from the solar wind; an extremely dense cloud-layer enveloped the planet between 22 and 30 miles above its surface and a cloud-top layer at 37.5 miles; a band of intense brightness appeared to ring the poles; there was thought to be an area of intense cooling of the atmosphere; polar vortex appeared to exist between the polar ring and the south pole, in which the cooled atmosphere sunk in a spiral flow to the surface maintaining the general circulation between the vortex and the equator, (It is assumed that there is also a northern vortex but the path of Mariner 10 did not permit that observation to be made); there was a region of relatively stable streamlike flow between the polar ring and the equator, within which the subsolar disturbance was located—the so-called Venusian eye which was probably a feature of its atmosphere rather than of its surface.

On March 16, 1974, the course of Mariner 10 was subjected to a final correction placing it in position for a fly-by of Mercury. The spacecraft began to transmit its clearest images of that planet on March 23, when 250 pictures were taken from a

The Planet Venus as Photographed by Mariner 10 from a Distance of 450,000 Miles (Source: NASA/JPL - Photo No. P-14400)

distance of 3.5 million miles. They revealed no details of the Mercurian surface, but were sharper than any taken by earth-based telescopic cameras. On the following day, additional pictures were transmitted showing the partial outline of the planet but no details of its surface.

Mariner 10 passed within 460 miles of Mercury, its closest approach on March 29. The pictures taken at that time, revealed a crater surface, pocked by meteorites and the effects of 4.6 billion years of close proximity to the sun. Traversing its surface were valley-like configurations of unknown origin.

The approach pictures did not reveal that the Mercurian surface had dry seas such as the moon but those transmitted when the spacecraft was receding from Mercury indicated the possibilities of such sea-like regions, although much of the surface appeared to be rough. The floors of some of the craters were completely smooth, except where pocked by the impacts of the meteorites, other than those which produced the craters originally. Many of the crater centers appeared to have central nipple-like peaks similar to those found in many craters on the moon. The white spot, appearing in the pictures of the central area of the region observed, was a small and very bright crater, 25 miles wide, cut into the wall of a larger, older, and darker crater. A system of rays radiating from the crater, appeared to consist of debris scattered about by the impact of the meteorites that had formed the crater. There were also many small bright craters scattered about the surface, assumed to be of more recent origin, and in one picture a 100-mile ridge was observable.

An unexpected revelation was that Mercury possesses a magnetic field about one per cent as strong as that of the earth and stronger than those detected near Mars and the moon. It was previously assumed that Mercury did not have a magnetic field

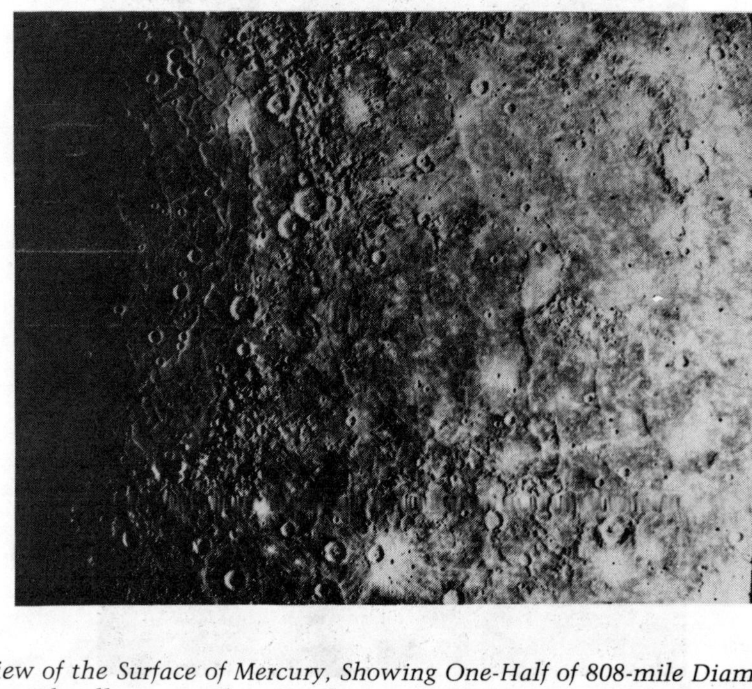

Close-up View of the Surface of Mercury, Showing One-Half of 808-mile Diameter Circular Basin. Hills and valleys extend outward in a radial manner from the main ring. (Source: NASA/JPL - Photo No. P-14678)

Mercury's Northern Limb. Shown is a prominent east-facing scarp extending from the limb near the middle of the photograph for hundreds of miles. This picture was taken by Mariner 10 at a distance of 49,000 miles from Mercury. (Source: NASA/JPL - Photo No. P-14679)

because of its slow rotation. It was also surprising to find that Mercury had an atmosphere containing helium, neon, argon with some hydrogen and little, if any, oxygen, carbon dioxide or nitrogen. *This was the first such planetary atmosphere ever to be discovered.*

As Mariner 10 passed beyond Mercury, it began a search lasting from March 31, through April 2, for possible Murcurian moons. For a time, it was thought that a small moon had been discovered but the object turned out to be the rim of a hot star known as 31 Crater in the Constellation Corvus.

As the spacecraft receded from Mercury, its temperature continued to rise because of an electrical malfunction, but its cameras continued to send back images of the planet. At the time, the heating problem raised serious doubts as to the feasibility of Mariner 10 making a second fly-by Mercury.

The data returned to earth by Mariner 10's sensoring devices relating to Mercury appear to support the theory that it, like other solar planets, developed a dense iron rich core below a less dense outer layer during its formative stages. This differentiation probably came about through the melting of its interior when the planet was formed. It also raises the possibility that the planet's exterior layer may consist of rock with high content of silicate.

The television images of the surface of Mercury indicated that it was very rough and covered with craters. There were many large basin-like features appearing to have been flooded with lava. One of these basins called Caloris, located just north of the planet's equator, appeared to be eight hundred miles wide, and is believed to be the hottest spot on Mercury as well as in the solar system. Another crater, named Kuiper, about twenty-five miles wide, was located in the northern hemisphere of the planet. It is believed that no significant alterations have taken place in the surface of Mercury since its early formative stages.

Mariner 10, responded on July 2, 1974, to a course change required for a second sweep by Mercury and on September 21, 1974 passed within 30,000 miles of it. After completion of this second fly-by, the spacecraft reentered solar orbit and it was hoped that it might again fly by Mercury 176 days later. This was contingent upon its fuel supply holding out and its instruments remaining operative.[6]

The success of Mariner 10's mission rests in its discovery of Mercury's magnetic field and the transmission of some 1200 pictures with a resolution of objects as small as 300 feet revealing numerous characteristics of Mercury and its environment. Because of this mission, there has been increased interest in sending other Mariner-type spacecraft to the vicinity of that planet before the end of the decade.

REFERENCES

1. The basic information for this chapter was derived from leading newspaper releases, including *The New York Times* and *Aviation Week and Space Technology*.
2. The solar energy of Jupiter is only four per cent of that received at the earth and because of this the spacecraft was equipped with four radioisotope thermonuclear unit generators which turn the heat from decaying radioactive plutonium into electricity to operate its systems.
3. This marks the boundary where the electrically charged particles of the solar wind meet the outer limits of the magnetic field of Jupiter.
4. The malfunctioning of the onboard imaging system may have been caused by radiation bombardment or some random force. It was estimated that the radiation levels were a thousand times greater than required to kill a man. The intensity was reached about ninety minutes before the spacecraft reached its nearest approach to Jupiter.
5. "By Jove, It's Hydrogen," *Time,* September 16, 1974, p. 82.
6. On Saturday, March 15, 1975, Mariner 10, its technical problems having been corrected, sent back good television pictures of Mercury as a prelude to its approach to the planet, when it will swoop to within 130 miles of the surface. It is anticipated that the photographs will be the best ever taken of any space body other than the earth and moon, and will show details of areas 50 yards in diameter.

Chapter VII

MANNED SPACE FLIGHTS OF THE MERCURY AND GEMINI MISSIONS

> *The Congress hereby declares that it is the policy of the United States that activities in space should be devoted to peaceful purposes for the benefit of all mankind.*
>
> *The Congress of the United States of America*[1]

The United States followed an orderly development in its efforts to land men on the moon. Its major space projects, directed to the accomplishment of manned flights to the moon and its vicinity, were Mercury and Gemini. These were concerned primarily with placing manned spacecraft in suborbital and orbital flights about the earth and returning them and the astronauts safely to earth.[2]

Mercury Project

The Mercury project was organized formally on October 5, 1958 under the sponsorship of the National Aeronautics and Space Administration. Its objectives were to place in earth orbits manned spacecraft, investigate man's capabilities in space, and safely recover astronauts and capsules from space.

The Mercury capsule had a nickel-cobalt outer shell and a titanium inner shell separated by insulation as a protection against heat, cold, and noise. It was equipped with communication and navigational instruments, devices to provide oxygen in and remove carbon dioxide from the air inside the capsule, altitude control jets, an ablation heat shield for protection of the capsule on reentry into the earth's atmosphere, three solid-fuel retrograde rockets for space maneuvers, and main and emergency parachutes. The specially designed reclining contour couch made it possible for astronauts to withstand increases in pressure at lift-off of the rocket and deceleration upon its reentry into the earth's atmosphere.[3]

Mercury capsule MA-1 was launched on July 29, 1960, having as its primary purpose the testing of the capsule under maximum airloads and afterbody heating rates under conditions of reentry. The mission proved unsuccessful because of a malfunction of the Atlas booster rocket system.

The capsule MR-1[4] was successfully fired into space on December 19, 1960, on a capsule-qualification test. The capsule reached a height of 135 miles and traveled 240 miles down range at speeds up to 4,300 miles an hour. After a flight of 16 minutes, the Mercury capsule landed off Grand Bahama Island and was recovered by a helicopter. It

The Mercury-Redstone Booster Rocket (Source: NASA - Photo No. 60-MRI-9)

landed within eight miles of the planned impact area with all systems functioning satisfactorily.

The MR-2 was launched on January 31, 1961, on a 16 minute suborbital flight, carrying a fully operational Mercury capsule, inside of which was a 37 1/2 pound live chimpanzee named Ham. The spacecraft rose faster and higher than anticipated because of a malfunction of the booster rocket, but fortunately the capsule's systems performed satisfactorily and the chimpanzee survived with no ill effects. It attained an altitude of 157 miles and traveled 418 miles down range and the capsule with Ham inside made a successful reentry, landed on the ocean, and was recovered by a helicopter about three hours after lift-off.

The MA-2[5] was launched on February 21, 1961 to test the escape system, the structural dependability of the Mercury capsule, and simulate an abort. The flight was successful and was a step forward in the direction of manned earth-orbital flights. The capsule separated from the booster rocket and coasted to an attitude of 107 miles. After about eighteen minutes, it landed in the Atlantic Ocean 1,425 miles down range having attained a maximum velocity of 12,850 miles an hour.

On March 18, 1961, a test capsule was sent into space to determine the reliability of the capsule's escape system and structure under abort conditions. The escape sequence malfunctioned causing the escape rocket to fire 14 seconds too soon. The capsule separated from the booster rocket 23 seconds later and tumbled away from it.

This test shot was followed on March 24, 1961, by a successful firing of a modified Redstone rocket on top of which was a boilerplate Mercury capsule. It attained an altitude of 115 miles and traveled a distance of 311 miles down range from the launching site. All of the tests were completed on this mission.

A 78-pound Mercury capsule MA-3 containing a mechanical astronaut-simulator was launched on April 25, 1961, in an attempt to place it into an earth orbit. The purpose of the mission was to test the capsule's systems and the worldwide communication, tracking, and recovery networks. The Atlas booster rocket failed to follow the programmed flight path and 40 seconds after lift-off a signal was sent by ground control starting the abort sequence, which took place at 14,000 feet. After an automatic delay of three seconds the rocket was destroyed. The capsule lifted clear of the booster rocket and attained a height of 24,000 feet and later landed undamaged 600 feet offshore. During its 60-hour flight, information was transmitted relating to interplanetary magnetic fields near the earth.

The first manned suborbital flight, MR-3, launched on May 5, 1961, with Astronaut Alan B. Shepard[6] at the controls, lasted for 15 minutes and 22 seconds. The Redstone carrier rocket produced a thrust of 78,000 pounds putting the 2,700-pound Mercury capsule, Freedom 7, into an earth suborbital path with a maximum altitude of about one hundred and sixteen miles at a top speed of 5,180 miles an hour. The capsule landed 302 miles down range and Astronaut Shepard was rescued and the capsule recovered. Alan Shepard had experienced 5 minutes, 4 seconds of weightlessness.

A second manned suborbital flight, MR-4, took place on July 21, 1961, after being delayed twice by unfavorable weather. The Mercury capsule, Liberty Bell 7, with Astronaut Virgil I. Grissom[7] at the controls, attained an altitude of 118 miles and traveled 303 miles down range. During his journey, Astronaut Grissom experienced five minutes of weightlessness and observed the booster separation, the jettisoning of the retrorocket, and the openings of the drogue and main parachutes through the large window of the space capsule. He operated the controls manually for a time, maintaining the proper attitude of the spacecraft and triggering the ignition of the retrorockets during the 22-second rocket firing in preparation for reentry. He was subjected to a maximum deceleration of 11 g's during the reentry into the earth's atmosphere and descent which he withstood without difficulty. The capsule, with the astronauts inside, landed in the Atlantic Ocean northeast of the Grand Bahama Island within sight of the recovery ship. Astronaut Grissom was forced to abandon the capsule before it was hooked to the recovery helicopter because of the premature explosion of the hatch bolts. He was in the water for four minutes before being rescued by helicopter and flown to the recovery ship USS *Randolph*. Unfortunately, the Mercury capsule sank in deep water and was not recovered. The mission was successful in demonstrating that the Mercury capsule was qualified for orbital flights.

The MA-4 mission, carrying an unmanned Mercury capsule, was launched on September 13, 1961, with the aid of an Atlas carrier rocket producing 360,000 pounds of thrust. The capsule was equipped with a mechanically simulated respiratory device, voice tapes to communicate with the Ground Tracking Station, and a fully operational attitude control system. Two minor malfunctions occurring during the flight were remedied by ground control but could have been corrected had there been an astronaut aboard. The mission lasted one earth orbit and after a successful reentry, the Mercury capsule landed in the Atlantic Ocean about one hundred and sixty miles east of Bermuda where it was recovered by the USS *Decatur*. The mission was successful in testing the effectiveness of the life-support systems, the worldwide tracking network, and the spacecraft's controls.

Another unmanned Mercury capsule MA-5 was sent on a two earth-orbital mission on November 29, 1961, carrying a five and a half year old male chimpanzee, Enos, weighing 37 1/2 pounds. Difficulties were encountered during the first orbit but all systems functioned satisfactorily and Enos carried out his four major tasks, the moving of several levers in sequence in response to flashing colored lights. During the second orbit, the capsule began to roll but was stabilized by jet thrusters resulting in more fuel being consumed than planned. In addition, the cooling system malfunctioned, causing the length of the mission to be shortened. As it approached the end of its second orbit, the capsule, containing the chimpanzee made a successful reentry and landed in the Atlantic Ocean at the planned area. Both capsule and chimpanzee were recovered by the USS *Stromes* and taken to Bermuda. Later, Enos, who was flown to Cape Canaveral for examination, showed no ill-effects from his venture into space.

John H. Glenn, Jr.[8] was carried into space on the MA-6 mission in the Mercury spacecraft, Friendship 7 on February 20, 1962. *This was the first manned earth-orbital flight achieved by an American astronaut.* The space capsule completed three full earth orbits with an apogee of 141 miles and a perigee of 86 miles. During 4.6 hours, Astronaut Glenn was subjected to weightlessness with no ill-effects. He took photographs of the earth, the horizon, clouds, and stars, and made visual observations of natural phenomena in space. The most interesting observation

The "Friendship" Space Capsule Mated to the Atlas Booster Rocket Showing Escape Tower (Source: NASA - Photo No. 62-MA-74)

was of luminous particles traveling with the spacecraft at every sunrise. A critical situation developed with the landing bag and the heat shield, forcing him, after the third full orbit, to operate the controls manually on reentry with the retro-package still attached to the heat shield. John Glenn achieved a successful reentry into the earth's atmosphere and splashed down in the planned area 700 miles southeast of Cape Kennedy within five miles of the recovery destroyer USS *Noa.* The major objectives of the mission to investigate man's capabilities in space and test the spacecraft and its supporting systems were successfully accomplished.

Astronaut Malcolm Scott Carpenter[9] on the MA-7 mission was the second American to complete an earth-orbital flight. The Mercury spacecraft, Aurora 7, which carried him into space, was launched on May 24, 1962, without a single hold in the countdown, into earth orbit with an apogee of 145 miles and a perigee of 86 miles. Scott Carpenter experienced about five hours of weightlessness during the three-orbital flight, which was slightly longer than that experienced by John Glenn. During the flight, he made visual observations and took some two hundred colored pictures of clouds, terrain, sunset, and the booster rocket. Of special interest were the sunrise particles "fireflies", the illuminated particles which John Glenn reported seeing during his mission. Just prior to reentry, he reported difficulty with the automatic control system in establishing the attitude for retrofire. This forced him to resort to manual-controlled retrofire on reentry, causing him to overshoot the planned recovery area by about two hundred and fifty miles. Astronaut Carpenter was sighted by search planes about an hour after splashdown and three hours later was rescued by a helicopter from the USS *Intrepid*. The capsule was recovered by the USS *Pierce* within six hours after splashdown.

The third manned earth-orbital flight MA-8 was initiated on October 3, 1962, when the Mercury capsule, Sigma 7, carrying Astronaut Walter Schirra[10] was fired into space and assumed an earth orbit with an apogee of 175 miles and a perigee of 99 miles. Astronaut Schirra experienced weightlessness for a period of 8 hours and 30 minutes with no apparent effects upon his ability to carry out his assigned experiments or follow the flight plan. This was remarkable because he had to work in an overheated spacesuit during the first orbit and a half of the flight. After six full orbits (9 hours and 13 minutes in space), he began his reentry and descent procedures and later splashed down in the Pacific Ocean near Midway Island about four miles from the aircraft carrier *Kearsarge,* from which location he was picked up within forty minutes. The mission proved to be one of the most successful in the Mercury series because it demonstrated the feasibility of future manned long-duration space flights.

Astronaut L. Gordon Cooper, Jr.[11] rode into space in the modified Mercury capsule, Faith 7, on May 15, 1963, for a one-day mission, designated as MA-9, to test the effects of extended weightlessness on man's physical condition and his ability to perform tasks. The flight lasted 34 hours and 20 minutes during which time the capsule completed 22.9 earth orbits. Several scientific and engineering experiments were carried out during this mission, among them aeromedical studies, radiation measurements, photographic studies, two visibility experiments, temperature measurements of the spacecraft and communication tests, including a television system.

Little difficulty was encountered until the 18th orbit, when the automatic attitude control failed, forcing Gordon Cooper to resort to manual control for the rest of the flight to ensure proper attitude of the capsule on reentry into the earth's

The "Sigma 7" Space Capsule Being Prepared for Mating with the Atlas D Booster Rocket (Source: NASA - Photo No. 62-MA8-80)

atmosphere. Astronaut Cooper was assisted verbally in the manual execution of the reentry procedures by John Glenn onboard the tracking ship, *Coastal Sentry Quebec*, stationed near the coast of Japan. He splashed down near Midway Island in the Pacific Ocean within four miles of the planned landing area and was rescued within thirty-five minutes by the USS *Kearsarge*. Astronaut Cooper remained in the capsule until it was hoisted aboard the recovery ship.

Gemini Project

The Gemini Project, an extension of the Mercury Project, was the second phase of the space program of the United States having as its goal the landing of men on the moon. The objectives of the project were to determine how men perform in space on long-duration flights, develop the capability to rendezvous and dock with other spacecraft, perform extravehicular activities, provide a basis for scientific, engineering, technological and medical experiments, develop methods of controlling the spacecraft's reentry flight paths, select landing areas, develop operational space flight experience, and return the astronauts and the spacecraft to earth safely.

The Gemini spacecraft was designed to hold two astronauts and had two major components, the adapter and the reentry module. The adapter was constructed in two parts: the equipment section and the retrograde section. The former was to be ejected in space when it was no longer needed and the latter was to be retained until reentry in order to reduce the speed of the module to a level at which it was safe for it to reenter the earth's atmosphere.

On April 8, 1964, the flight stage of the Gemini Project was initiated when unmanned Gemini I was launched by a Titan II launch vehicle on a flight test. The partially equipped configuration of the reentry module was sent on a 65 earth-orbital flight, making it *the second heaviest American spacecraft placed in an earth orbit* exceeded only by a spacecraft launched by a Saturn I on January 29, 1964.

The second unmanned Gemini flight was slow in getting off the pad, because of delays. The carrier rocket, with the Gemini capsule on top, was positioned on the pad and ready for launching, when it was struck by lightning and had to be removed from the pad to avoid damage from hurricanes Cleo and Dora. The launching was rescheduled for December 8, 1964, but the lift-off had to be postponed again because of a rocket engine malfunction. Finally, on January 19, 1965, Gemini II was sent on a suborbital flight after a minor delay of 62 minutes, caused by the spacecraft. On this mission, the Gemini capsule proved trustworthy in that all of the planned experiments were completed and its control systems performed in a satisfactory manner. The mission served as the final flight qualification prior to manned orbital flights. The final countdown was not interrupted by carrier-vehicular delays and only one minor delay occurred because of a malfunction with respect to the spacecraft. The primary objectives of this mission were to demonstrate the structural integrity of the space capsule and verify the reentry heat procedure under the most severe conditions anticipated upon reentry. The recovery of the spacecraft was achieved by the USS *Lake Champlain*, 2,125 miles southeast of Cape Kennedy.

The Gemini III mission, the first of two scheduled manned flights in the Gemini program, was launched on March 23, 1965. Astronaut Virgil I. Grissom,[12] the command pilot, and John W. Young,[13] the pilot, spent 4 hours and 53 minutes orbiting the earth on a path with an apogee of 140 miles and a perigee of 100 miles. The objectives of the mission were to demonstrate precise orbital maneuvering, and evaluate the capabilities of the astronauts to carry out manual control of the spacecraft in orbital flight, including their ability to withstand the conditions of extended time in space. In addition, they were to test the worldwide ground-tracking network, the pre-launch and launch procedures, and the recovery systems and procedures. *This marked the first time that an orbital path of a spacecraft had been changed in flight by simulating a rendezvous and docking maneuver*. The astronauts and the space capsule were recovered by the aircraft carrier USS *Intrepid*, after splashing down in the Atlantic Ocean about four hundred and eighty miles southwest of Bermuda.

The launching of Gemini IV was achieved on June 3, 1965, when astronauts James A. McDivitt,[14] command pilot, and Edward H. White,[15] pilot, were placed in an earth orbit with an apogee of 175 miles and a perigee of 100 miles. The mission had two major purposes: to evaluate effects upon the astronauts of prolonged exposure to the spatial environment and the performance of the spacecraft's control system during an extended flight in space. There were six secondary purposes and 11 experiments set for the mission: three were medical, four engineering, two technological, and two scientific, all of which were completed satisfactorily.

The duration of the mission was 97 hours and 56 minutes of which 22 minutes were spent outside the capsule by Astronaut White in extravehicular activities (EVA), with a propulsion unit in his hand to assist him in his maneuvers. He remained at all times attached to the capsule by an umbilical cord, without which he would have floated off into space and to a fiery destruction upon reentry into the earth's atmosphere. *He was the first American to walk in space and take the first external*

Astronaut Edward H. White Carrying Out Extravehicular Activities During the 3rd Orbit of the Earth by Gemini IV (Source: NASA - Photo No. 65-HC-361)

pictures of a spacecraft in flight during extravehicular activities. The astronauts and the capsule landed 630 miles southwest of Bermuda and were recovered by the USS *Wasp* on June 7.

On August 21, 1965, Gemini V was placed in an earth orbit carrying Astronauts L. Gordon Cooper,[16] command pilot, and Charles Conrad, Jr.,[17] pilot, on a mission lasting eight days. The mission demonstrated the capacity of trained men to perform effectively in space for relatively long periods and the reliability of onboard fuel cells as a source of power for operating instruments. The mileage covered by Gemini V, in orbiting the earth, was equivalent to that of a roundtrip to the moon. The flight was successful and the astronauts were returned to earth safely.

Efforts were made to launch Gemini VI on October 25, 1965, but the lift-off was aborted because the Agenda vehicle engine malfunctioned. As a result Gemini VII, in accordance with an alternate plan, was placed in orbit on December 4, 1965, before Gemini VI, and was actually the fourth manned flight in the Gemini Program. Aboard were astronauts Frank Borman[18] and James A. Lovell,[19] who spent 13 days, 8 hours and 35 minutes in space. It is interesting to note, that they did not wear the usual spacesuits but were clothed in light-weight pressure suits which permitted them greater freedom of movement and more comfort. They even removed the suits for a time during the flight, and moved about the cabin in their shirt sleeves. They were successful in carrying out 18 of the 20 planned experiments during the flight.

The Salton Sea and the Imperial Valley in Southern California as Seen from the Earth-Orbiting Gemini V Spacecraft (Source: NASA - Photo No. 65-H-1374)

The launching of Gemini VI-A, formerly designated Gemini VI, was undertaken on December 12, 1965, but was aborted within 1.19 seconds after ignition because of a malfunction of a lift-off tailplug. After the carrier vehicle, pad, and umbilical tower were checked and/or repaired, the spacecraft was launched on December 15, 1965, 11 days after the launching of Gemini VII. It carried astronauts Walter M. Schirra, Jr.,[20] and Thomas P. Stafford[21] into space for the purpose of a rendezvous with Gemini VII space capsule already in an earth orbit. *The first successful rendezvous by two spacecraft was accomplished within 5 hours and 56 minutes after lift-off.* Gemini VI-A was maneuvered to within one hundred and twenty feet of Gemini VII and for three and a half orbits, the two spacecraft carried out station-keeping maneuvers at distances varying from one foot to 300 feet from each other. Both capsules splashed down safely, landing close to the planned landing areas and were recovered along with the astronauts.

Gemini VIII, the sixth manned flight in the project, succeeded in effecting a rendezvous with a target vehicle and docked with it while in earth-orbital flight. The unmanned Agena target vehicle was launched on March 16, 1966, followed by Gemini VIII 1 hour and 41 minutes later with astronauts Neil A. Armstrong,[22] command pilot and David R. Scott,[23] pilot. During the fourth orbit, Gemini VIII succeeded in making a rendezvous and docking with the Agena-target vehicle. The rendezvous phase was completed after 5 hours and 58 minutes of Gemini VIII's flight, and nine maneuvers, bringing it within one hundred and fifty feet of the target vehicle. Station-keeping activities were carried on for 36 minutes, after which Gemini VIII docked with the target vehicle. The linked spacecraft exhibited stability and control for about twenty-seven minutes, after which there developed considerable roll and yaw. Even after the astronauts had undocked the spacecraft from the target vehicle, it still continued to roll and yaw. The flight, following a controlled reentry, was terminated and the capsule and the crew were safely returned to earth, landing within seven miles of the planned landing site in the Western Pacific Ocean. The space capsule and the astronauts were recovered by the USS *Leonard Mason* about 3 hours and 11 minutes after splashdown. The Agena-target vehicle, after 11 maneuvers was placed in a 200-mile circular earth-parking orbit to be used again, if the need should arise.

The launching of Gemini IX on a mission involving extravehicular activities (EVA), was attempted on May 17, 1966. The Atlas-Agena target vehicle lifted off but a malfunction resulted in an erratic trajectory and its loss. The mission was rescheduled, and on June 1, 1966, an augmented target-docking adapter was launched into a circular earth orbit of 161 miles. Gemini IX-A, formerly designated Gemini IX, was to follow but the countdown was stopped at T minus three minutes. The launching was rescheduled for June 3, 1966, at which time, Thomas P. Stafford,[24] command pilot, and Eugene A. Cernan,[25] pilot, were hurled into an earth orbit. They made several rendezvouses with the target vehicle but were unable to dock with it because the shroud, which protected it from aerodynamic forces and heat during the launch, was still partially attached.

On the third day of the Gemini IX-A flight, Astronaut Cernan carried out extravehicular activities for 2 hours and 5 minutes, during which time he checked out an astronaut-maneuvering unit and performed difficult maneuvers, while attached to the capsule by a 100 foot umbilical cord. The extravehicular activities caused him to tire and the visor of his helmet became fogged reducing his visibility. This made it necessary for him to terminate the extravehicular activities and return to the cabin of the capsule. A controlled reentry was achieved and the spacecraft with the astronauts, landed within 2.8 miles of the selected point, which was the best executed controlled reentry of any mission up to that time.

Gemini X spacecraft, with John W. Young,[26] command pilot and Michael Collins,[27] pilot, was launched into an earth orbit on July 18, 1966. The purposes of the mission were to rendezvous and dock with an Agena target vehicle, test the propulsion systems, practice docking, and carry out extravehicular activities. The last objective was not accomplished. The target vehicle was launched shortly before Gemini X and within 5 hours and 21 minutes the rendezvous was achieved and 31 minutes later docking was also accomplished. The linked spacecraft orbited the earth for almost 38 hours and 47 minutes in an orbit with an apogee of 476 miles and a perigee of 182 miles. Twenty-three hours and 24 minutes after the combined flight started, the hatch of Gemini X was opened for 49 minutes, during which time Astronaut Collins performed standing extravehicular activities which were ended when both astronauts began to experience eye irritation.

Collins emerged again from the spacecraft, 48 hours and 41 minutes after lift-off and spent 39 minutes in space attached to it by an umbilical cord. While in space, he retrieved an experimental package which had been attached to the target vehicle since March. After undocking and a successful reentry, the mission was terminated on July 21, with the splashdown of the Gemini X capsule in the Atlantic Ocean about

Circling the Earth, an Agena for Use as the Target Vehicle in the Gemini Program (Source: Lockheed Missiles and Space Company)

Astronaut Aldrin Performing Extravehicular Activities During Gemini XII Flight (Source: NASA - Photo No. 66-HC-1939)

three and a half miles off the planned landing area. The astronauts were picked up by helicopter and flown to the USS *Guadalcanal*, 27 minutes after splashdown and in another 27 minutes the capsule was recovered.

On September 12, 1966, astronauts Charles Conrad, Jr.,[28] command pilot, and Richard F. Gordon,[29] pilot of Gemini XI, were placed in an earth orbit for the purpose of continuing space maneuvers and experiments. The rendezvous and docking were accomplished with the unmanned target vehicle during the first orbit. *The astronauts attained the highest altitude, up to that time, in a manned space capsule while docked with the Agena target vehicle.* Using the power of the target vehicle, they maneuvered the linked vehicles into an earth orbit with an apogee of 853 miles and a perigee of 178 miles.

Astronaut Gordon conducted extravehicular activities for approximately half an hour, during which time he tethered the target vehicle to Gemini XI. This required the expenditure of considerable energy on his part, resulting in the extravehicular activities being terminated after 33 minutes. On the third day in space, Gemini XI was undocked from the Agena target vehicle and a tethered formation was maintained for about two orbits. *The first rendezvous, using onboard computation, was achieved* as well as two practice dockings in which both astronauts participated. *It was on this mission, that reentry into the earth's atmosphere was totally performed by a computer for the first time.* Retrofire occurred after an elapsed time of 70 hours, 41 minutes and 36 seconds in flight and splashdown took place 35 minutes later. Gemini XI touched down in the Pacific Ocean 2.7 miles from the planned landing area and the astronauts were picked up by helicopter and flown to the recovery ship the USS *Guam* 24 minutes after splashdown on September 15, and the spacecraft was recovered 35 minutes later.

Gemini XII, the last earth-orbital mission of the Gemini Project, was initiated on November 11, 1966, carrying Astronauts James A. Lovell, Jr.[30] and Edwin E. Aldrin, Jr.,[31] into space, for the purpose of performing experiments, extravehicular activities, and rendezvousing and docking with a target vehicle, all of which were accomplished. *During the flight, Astronaut Lovell set a record having logged a total of 425 hours, 10 minutes and 2 seconds in space. Astronaut Aldrin also set a record having logged a total of 5 hours and 26 minutes in extravehicular activies.* Retrofire took place after 93 hours, 59 minutes and 58 seconds of the flight and splashdown was achieved 34 minutes and 33 seconds later. The mission was successfully terminated with the splashdown of the spacecraft in the Pacific Ocean 2.6 miles off the planned landing area and the astronauts were taken aboard a helicopter and flown to the USS *Wasp* on November 15, 30 minutes after splashdown. The space capsule was recovered 1 hour and 7 minutes later.

REFERENCES

1. Public Law 85-568, 85th Congress, National Aeronautics Act of 1958, July 29, 1958, Section 102(a).
2. The basic information for this chapter was obtained from the *Semiannual Reports of the National Aeronautics and Space Administration*, newspaper releases, including *The New York Times* and *Aviation Week and Space Technology*.
3. In addition, the astronaut wore a pressurized spacesuit to protect him further from the hazards and discomfortures of the space flight.
4. A combination of the Redstone Rocket and the Mercury Capsule.
5. A combination of the Atlas Booster Rocket and the Mercury Capsule.
6. *Apollo 14 Flight*, January 31, 1971 (p. 113).
7. *Gemini Flight III*, March 23, 1965 (p. 83). Astronaut Grissom lost his life in the Apollo Cabin fire on January 27, 1967 (p. 92).
8. Made only one flight into space.
9. Made only one flight into space.
10. *Gemini Flight VI-A*, December 12, 1965 (p. 85), and *Apollo 7 Flight*, October 11, 1968 (p. 94).
11. *Gemini V Flight*, August 21, 1965 (p. 84).
12. *Mercury Flight*, Liberty Bell 7, July 21, 1961 (p. 79).
13. *Apollo 10 Flight*, May 18, 1969 (p. 97).
14. *Apollo 9 Flight*, March 3, 1969 (p. 96).
15. Astronaut White lost his life in the Apollo Cabin Fire on January 27, 1967 (p. 92).
16. *Mercury Flight, Faith 7*, May 15, 1963 (p. 81).
17. *Gemini XI Flight*, September 12, 1966 (p. 88).
18. *Apollo 8 Flight*, December 21, 1968 (p. 96).
19. *Gemini XII Flight*, November 11, 1966 (p. 88); *Apollo 8 Flight*, December 21, 1968 (p. 96); and *Apollo 13 Flight*, April 11, 1970 (p. 109).
20. *Mercury Flight, Sigma 7*, October 3, 1962 (p. 81), and *Apollo 7 Flight*, October 11, 1968 (p. 94).
21. *Gemini IX-A Flight*, June 3, 1966 (p. 86), and *Apollo 10 Flight*, May 18, 1969 (p. 97).
22. *Apollo 11 Flight*, July 16, 1969 (p. 99).
23. *Apollo 9 Flight*, March 3, 1969 (p. 96), and *Apollo 15 Flight*, July 26, 1971 (p. 118).
24. *Gemini VI-A Flight*, December 15, 1965 (p. 85), and *Apollo 10 Flight*, May 18, 1969 (p. 97).
25. *Apollo 10 Flight*, May 18, 1969 (p. 97).
26. *Gemini III Flight*, March 23, 1965 (p. 83), and *Apollo 10 Flight*, May 18, 1969 (p. 97).
27. *Apollo 11 Flight*, July 16, 1969 (p. 99).
28. *Gemini V Flight*, August 21, 1965 (p. 84), and *Apollo 12 Flight*, November 14, 1969 (p. 104).
29. *Apollo 12 Flight*, November 14, 1969 (p. 104).
30. *Gemini VII Flight*, December 4, 1965 (p. 84), and *Apollo 8 Flight*, December 21, 1968 (p. 96).
31. *Apollo 11 Flight*, July 16, 1969 (p. 99).

Chapter VIII

MAN'S FIRST LANDING ON THE MOON

"Tranquility Base here: The Eagle has landed."

****

"That's one small step for a man, one giant leap for mankind."

Astronaut Neil Armstrong[1]
July 20, 1969

The Apollo Project covered almost a decade of careful planning and testing. The years 1961 and 1962 were given primarily to decision-making relating to its developmental aspects and during 1963, contracts were let for various components and the designing and manufacturing of facilities were undertaken. During 1964 and 1965, there followed further designing, manufacturing, and testing, leading up to the start of the project in 1966 and the landing of men on the moon by 1969. The project was the most ambitious space undertaking of the United States which was made possible through the accumulation of information and experience from the Mercury and Gemini projects as well as scientific meteorological and communication programs.[2]

The three-man Apollo spacecraft consisted of three basic components each adapted for a particular phase of a lunar mission: earth orbiting, circumlunar orbiting, and lunar landing and lift-off. The command module served as the cabin for the astronauts during the launch and reentry phases of a lunar mission and contained the flight control and transmitting facilities. The service module, containing the power and central guidance systems, played important roles in several phases of the Apollo space flight. In earth orbital flight, it remained with the command module while in space and could be used in orbital rendezvous and other maneuvers. In lunar flight it remained with the command module and continued in the circumlunar parking orbit waiting for the return of the astronauts in the lunar-excursion module. The lunar-excursion module, consisting of a descent and an ascent section, was designed to land astronauts on the surface of the moon and to lift them off and return them to the command-service module in the lunar-parking orbit. Only the command module was designed to reenter the earth's atmosphere and to be recovered on the earth's surface.

A lunar mission is initiated by firing the Saturn V carrier rocket and after the space unit has attained a given altitude, the booster stage of the carrier rocket separates leaving the other stages and the Apollo capsule to continue their ascent. The launch escape tower is jettisoned and the second stage of the carrier rocket ignites imparting to the third stage, housing the lunar-excursion module, and the command-service module, an additional thrust which inserts the unit into an earth-parking orbit. Later, the third stage ignites, injecting the unit into a lunar trajectory, and after a short interval, the third stage, together with the lunar-excursion module, separates from the command-service module. The command-service module then pivots and eases up to and links with the lunar-excursion module. After docking is achieved, the third stage is jettisoned and enters a solar orbit, leaving the command-service module and the lunar-excursion module, as a unit, to proceed on a lunar trajectory.

The spacecraft is subjected to three midcourse corrections, if necessary, after which the astronauts inspect the lunar-excursion module before the spacecraft is injected into a lunar-parking orbit. After the spacecraft has assumed a lunar-parking orbit and has made about two revolutions, the lunar-excursion module, with two astronauts inside, is placed in a more circular lunar orbit from which it can descend to the surface of the moon. If it is to land on the moon, its speed is reduced by firing the descent-section engine, which enables it to soft-land.

The Apollo Command-Service Module (Source: NASA - Photo No. 69-H-1791)

On lift-off from the moon's surface, the astronauts reenter the ascent section of the lunar-excursion module and fire its engine, causing it to ascend and leave behind the descent section which serves as a launching platform. After the ascent stage has rendezvoused and docked with the command-service module in the lunar-parking orbit, the astronauts perform station-keeping duties and transfer the cargo to the command-service module. On completion of the transfer, they enter the command-service module and secure the hatches, after which the ascent section of the lunar-excursion module is jettisoned. The command-service module continues to circle the moon in the lunar-parking orbit and when it is on the far side of the moon, its engine is fired inserting it into an earth trajectory, which is adjusted, if necessary, by midcourse corrections. Before reentering the earth's atmosphere, the service module is jettisoned and the command module is turned so that its blunt end, with the heat shield, faces in the direction of its movement. The command module slows down upon reentering the earth's atmosphere and when near the earth's surface, parachutes are deployed allowing it to descend slowly to the surface of the earth.

The first phase of the Saturn I launch vehicle program was concluded with the test flight of the rocket on July 30, 1965 and on February 26, 1966, a new phase was initiated in the Saturn-Apollo program, when the 132-foot liquid-fuel Saturn 1B carrier rocket hurled into space an unmanned 37,500-pound Apollo payload. Following a three-hour delay and a false start, the carrier rocket, with 1.6 million pounds of thrust at lift-off, sent the spacecraft on a 5,500-mile suborbital test flight lasting 39 minutes. The spacecraft having passed its preliminary tests satisfactorily was to be used to send crews of three astronauts to orbit and/or land on the moon. It splashed down 4,500 miles down range in the South Atlantic 200 miles southwest of Ascension Island, 38 miles short of its planned target area. Within two hours it was recovered by a helicopter from the USS *Boxer*.

Again on August 25, 1966 an unmanned Apollo spacecraft was sent on a successful 93-minute suborbital test flight of 18,000 miles. This test mission marked the third successful firing of the 22-story Saturn 1B carrier. It was the most powerful version of earlier Saturn rockets and the forerunner of the 7.5 million-pound-thrust Saturn V rocket. The spacecraft landed 200 miles short of its target in the Central Pacific from which it was recovered.

A tragic accident occurred on January 27, 1967, during the preparatory phase of the Apollo 1 manned earth-orbital 14-day mission scheduled for lift-off on February 21, 1967. Three astronauts lost their lives: Lieutenant Colonel Virgil I. Grissom, one of the seven original Mercury astronauts;[3] Lieutenant Colonel Edward H. White, the first American to walk in space,[4] and Lieutenant Commander Roger B. Chaffee, awaiting his first space flight. They were secured inside the Apollo capsule at the time of the accident carrying on a full-scale simulation of the launching procedures, when suddenly the inside of the capsule burst into flames. Every effort was made to save them but the intense heat, together with the fact that the hatch was closed and could not be opened in time, made their rescue impossible.

A Board of Inquiry was created to investigate the cause or causes of the fire and in April, 1967, the Board released a 3,000 page report of its findings.[5] The probable cause of the fire was ascribed to faulty electric wiring but it concluded that the exact cause or causes could never be ascertained positively. The irony of this fatal accident was that the astronauts, who had been so highly trained to cope with the dangers that they might encounter in space, died in a simulation exercise of the lift-off of their scheduled flight, without leaving the surface of the earth. *They were the first*

The First Saturn 1B to Successfully Launch an Unmanned Spacecraft on a 300-mile High Suborbital Flight (Source: NASA - Photo No. 66-H-120)

American astronauts to lose their lives while actually engaged in the space programs of the United States.

The accident slowed down the momentum of the Apollo Project, which had been accelerating over the years, and raised serious doubts as to the possibility of landing men on the moon before the end of the decade, the goal set by President John F. Kennedy in May 1961. Fortunately, the momentum was regained, giving renewed vigor and determination to the revised launching program which achieved successful launchings during the autumn of 1968.

Unmanned Apollo Test Space Flights

The unmanned Apollo 4 spacecraft was launched on November 9, 1967.[6] Its primary purposes were to test the Saturn V carrier rocket in combination with the Apollo spacecraft, the heat shield of the command module under actual reentry conditions, and the performances of all systems. It was a critical mission because it was the first time that such a huge and complex three-stage carrier vehicle had been tested in combination with the Apollo spacecraft in space on an all-up basis. The overall performance of the carrier rocket and the primary test objectives of the mission were accomplished and the space capsule was recovered less than five miles from the planned landing point.

The lunar-excursion module was tested on January 22, 1968, by the unmanned *Apollo 5* mission. After minor delays, a Saturn 1B carrier rocket put a modified lunar-excursion module with a "mechanical boy" in an earth orbit with an apogee of 104 miles and a perigee of 76 miles. Although a premature engine cut-off caused the ground controllers to modify the mission's plan, the primary objective of the mission was accomplished. The modified lunar-excursion module was not returned to earth, but was jettisoned before reentry of the spacecraft and was disintegrated by the intense frictional heat. However, the Apollo spacecraft was recovered after returning to earth.

On April 4, 1968, following several postponements of lift-off, the unmanned Apollo 6 spacecraft was placed in an earth orbit by a Saturn V carrier rocket. The primary goals of the mission were to test the structural and thermal capabilities and the reliability of the carrier rocket and the Apollo capsule. The capsule was hurled into an unplanned orbit, because two of the five engines of the carrier rocket's second stage engines shut down prematurely and the third stage engine failed to reignite. As a result the spacecraft reached an altitude of only 13,800 instead of 320,000 miles. The Apollo capsule splashed down in the Pacific Ocean northwest of Hawaii 50 miles from the planned landing site and was recovered by the *USS Okinawa*.

The tests of the Saturn V carrier rocket and the Apollo spacecraft were terminated on June 24, 1968, when Joseph Kerwin, Vance Brand, and Joe Engle completed successfully a mock space flight of 187 hours in an Apollo capsule at a simulated altitude of more than one hundred and ten miles. The next phase was the launching of the first manned Apollo earth-orbital mission, similar to that which had been scheduled for February 21, 1967, but had been abandoned because of the fatal fire of January 27, 1967.

Manned Apollo Space Flights

On October 11, 1968, the 32,000-pound *Apollo 7* spacecraft, carrying Captain Walter M. Schirra, Jr.,[7] commander, Major Donn F. Eisele, navigator, and R. Walter Cunningham, a civilian astronaut systems engineer, was hurled by a Saturn 1B carrier rocket on an 11-day mission into an elliptical earth orbit with an apogee of 183 miles and a perigee of 140 miles, traveling at a speed in excess of 17,000 miles an hour. The major purposes of the mission were to test the modified Apollo spacecraft's capability for transporting three astronauts to the moon and returning them to earth safely and to carry out critical maneuvers in space, essential to a successful lunar voyage. This was the first attempt to fly a manned spacecraft in an earth orbit after the fatal fire of January 27, 1967, or since the Gemini XII Flight of November 1966.

The spacecraft and the third stage of the Saturn 1B carrier rocket, measuring 113 feet in length and weighing 69,000 pounds, remained attached for approximately three hours. Then the command-service module separated from the third stage, pulled 50 feet ahead of it, turned around and faced it. It then approached within four or five feet of it and traveled in formation with it for about fifteen minutes before moving out in front by about eighty miles in a lower orbit. On the second day, during the 19th orbit, a rendezvous was made with the spent third stage, the spacecraft approaching as near as seventy feet. To adjust the orbit for rendezvous, the large engine of the service module was fired, which was the first time that it had been tested in space on a manned flight. Although difficulties were encountered, the mission, for all practical purposes, was successful in that more was accomplished than had been planned.

It was on this mission that Captain Schirra refused to carry out an order by the Ground Control Station to turn on the onboard television camera. He did this to conserve the spacecraft's power, prior to the rendezvous with the third stage of the carrier vehicle. After the rendezvous, the television camera was activated and a program lasting about ten minutes was transmitted to earth. *This was the first time that live television pictures, originating from an earth-orbiting spacecraft, were viewed by the public.*

The service module was jettisoned just before the reentry of the command module into the earth's atmosphere after which the blunt end of the command module with its heat shield was turned in line with its forward movement. The reentry was executed successfully and on October 22, the Apollo 7 capsule, with the three astronauts inside, parachuted to an inverted splashdown in the Atlantic Ocean about three hundred and twenty-five miles south of Bermuda. It was righted by its flotation balloons and the astronauts were picked up by the aircraft carrier USS *Essex.*

The mission involved a 4.5-million mile return journey, taking approximately two hundred and sixty hours, and was equivalent to going 163 times around the earth. It was the longest manned space flight since Gemini VII,[8] and created a wave of optimism, which proved influential in bringing about the "go" signal for the Apollo 8 mission, scheduled for launching during December of that year.

Contrast Between Nature and the Space Age — a lone egert in flight and the launching of Apollo 8 (Source: NASA - Photo No. 68-H-1352)

The first manned space flight to orbit the moon was initiated on December 21, 1968 when the spacecraft Apollo 8, carrying Colonel Frank Borman,[9] *commander, Captain James A. Lovell, Jr.,*[10] *navigator, and Major William A. Anders, systems engineer, was fired into space by a Saturn V carrier rocket.* The lunar-excursion module was eliminated on this mission and only the command-service module was placed in lunar orbit. The main objective of the mission was to demonstrate that the Apollo spacecraft possessed the capability of executing critical maneuvers, at high rates of speed, required to carry the astronauts to and from the moon. All maneuvers were accomplished satisfactorily and *the three astronauts became the first human beings to orbit the moon.*

The mission also was man's farthest venture into space. It was also the first time that he had traveled faster than 17,400 miles an hour in a near-earth orbit and had been inserted into a trajectory toward a celestial body at a speed in excess of 24,200 miles an hour. The enormous momentum of the spacecraft required to escape from the 118-mile earth orbit, after two revolutions, decreased as the spacecraft coasted away from the gravitational influence of the earth, which was being countered by the moon's gravitational influence. It dropped to about twenty-seven hundred miles an hour before entering the equigravisphere, where the two gravitational forces are neutralized, and then increased, reaching about fifty-six hundred miles an hour, as it descended toward the moon. The spacecraft's momentum of descent was reduced by firing its engine in order to place it in an elliptical lunar orbit with an apolune of 169.1 miles above the front or light side of the moon and a perilune of 60.5 miles above its far or dark side. *The astronauts were the first to see the rough and rugged farside of the moon and the first to be cut off from radio contact with the earth while in lunar orbit.* During each orbit of the moon, they were out of contact with ground control for about forty-five minutes while traversing its far or dark side. The spacecraft was placed in a more circular lunar orbit after completing two elliptical orbits bringing it within 70 to 69.5 miles of the moon's surface. *They were also the first to view a lunar sunrise which they described as a strange and awesome sight.*

The astronauts were in the lunar orbit for approximately twenty hours and on the morning of the day before Christmas transmitted the first live television program from the vicinity of the moon. Later that day, they transmitted their second and last televised program from lunar orbit. During the programs, they described the lunar features, read passages from the Bible, relating to the creation of the world, and extended holiday greetings. After orbiting the moon ten times, the main engine was fired, inserting the capsule into an earth trajectory. The six-day mission was terminated when the spacecraft, with its occupants, splashed down in the Pacific Ocean about four miles from the recovery ship the USS *Yorktown*. During their time in space, they took numerous pictures of the moon's surface, and made valuable lunar observations, useful for planning other manned Apollo missions that followed and ultimately led to landing men on the moon.

After being delayed for about three days in the lift-off of Apollo 9 spacecraft, because of the threat of a common cold virus, Astronaut James A. McDivitt,[11] commander of the mission, David R. Scott,[12] navigator, and Russell L. Schweickart, a civilian system engineer, were sent into space on March 3, 1969, on a ten-day earth-orbital critical test flight. The major objectives of the mission were to test, under actual space conditions, the capability and dependability of the lunar-excursion module, and to carry out critical maneuvers essential for landing men on the moon and returning them safely to earth. It was the first time that the lunar-excursion module had been put into a near-earth orbit.

The carrier rocket together with the spacecraft, consisting of the command-service module and the lunar-excursion module, housed in a metal section between the carrier rocket's third stage and the command-service module, were launched as a unit. Following the separation of the booster rocket, the third stage, with the spacecraft attached, assumed an orbit and circled the earth for a time, after which the third stage with the lunar-excursion module attached, was separated from the command-service module and the command-service module was then moved 50 feet back. This maneuver was followed by a nose-to-nose relinking of the two modules, which flew together to test the docking latches, and refiring of the service module's engine. The third stage was jettisoned into a solar orbit.

On March 6, Astronauts McDivitt and Schweickart entered the lunar-excursion module to inspect its systems and then returned to the command-service module. On the following day, they reentered it and Astronaut Schweickart opened the hatch and spent about forty minutes outside the lunar-excursion module testing with the aid of handrails the reliability of the spacesuit and the back-pack, and taking pictures. During this time, he was dependent solely on the life-support system of his back-pack to provide him with the necessary supply of oxygen, air conditioning and communication contacts. *This was the first time that an American astronaut had left an orbiting spacecraft and depended solely on his back-pack instead of an umbilical cord.*

The lunar-excursion module was separated on March 7 from the command-service module, piloted by Astronaut Scott, and was flown about a hundred miles away from it by the two astronauts. This maneuver, in the course of which the descent stage of the lunar-excursion module was jettisoned, was aided by radar and lasted for six and a half hours. The final relinking of the lunar-excursion module with the command-service module occurred on the afternoon of the same day after which the two astronauts in the lunar-excursion module transferred to the command-service module and the lunar-excursion module was jettisoned condemning it to destruction upon reentry into the earth's atmosphere. After 151 orbits, the service module was also jettisoned and the *Apollo 9* command module reentered the earth's atmosphere. On March 13, it splashed down in the Atlantic Ocean about three hundred miles north of Puerto Rico and less than a mile from the planned landing area, where the USS *Guadalcanal* was standing by to recover it and rescue the astronauts. In 48 minutes after splash-down, the astronauts were onboard the recovery ship.

This mission was the most complex to date in the lunar-landing program, and four space records were established: the placing of the heaviest payload in an earth orbit, weighing 94,964 pounds for the total unit consisting of the command-service and the lunar-excursion module; the longest duration flight of two linked space vehicles lasting for 89 hours and 37 minutes; the longest duration for a manned group flight lasting for 6 hours and 20 minutes and involving the command-service module piloted by Astronaut Scott and the lunar-excursion module flown by Astronauts McDivitt and Schweickart; and the greatest mass ever assembled in space equal to 91,058 pounds for the two linked spacecraft.

May 18, 1969 marked the launching of *Apollo 10* spacecraft with Colonel Thomas P. Stafford,[13] commander of the mission, Commanders John W. Young,[14] and Eugene A. Cernan,[15] pilot and system engineer, respectively. The mission was for the purpose of rehearsing the procedures for landing the lunar-excursion module on the moon and its safe return to earth. It was the first time that the lunar-excursion module, slightly modified because it was not to land on the moon, was to be tested under lunar orbit conditions with two astronauts inside.

After one and a half orbits, with an apogee of 118 miles and a perigee of 115 miles, the command-service module, with the third stage of the carrier vehicle housing the lunar-excursion module, was propelled at approximatey 24,200 miles an hour into a lunar trajectory. Shortly thereafter the command-service module was separated from the third stage, housing the lunar-excursion module, and was turned around and nose-linked with the lunar-excursion module. The third-stage rocket was jettisoned after docking had been accomplished, and the command-service module with the lunar-excursion module linked to it, continued toward the moon. On reaching the vicinity of the moon, the command-service module, with the lunar-excursion module attached, was turned so that the lunar-excursion module faced in the direction of the spacecraft's forward movement, and preparations began for inserting the space unit into a lunar-parking orbit. This was accomplished by firing the service module's engine, placing the unit in an elliptical orbit with an apolune of 196 miles and a perilune of 70 miles. The spacecraft remained in this orbit for four and a half hours after which the orbit was made more circular with an apolune of 70 miles and a perilune of 69 miles.

On May 22, Astronauts Stafford and Cernan entered the lunar-excursion module, leaving Astronaut Young in the command-service module. They maneuvered the lunar-excursion module into a lunar orbit which brought it within 9.4 miles of the moon's surface. *This was the closest that man had ever approached the surface of the moon.* From this distance, the astronauts were able to make more accurate observations, photograph the lunar features with more detail, and survey the selected landing sites in the Sea of Tranquility for the Apollo 11 flight scheduled for midsummer.

The descent section of the lunar-excursion module was jettisoned following the second orbit and the engine of the ascent section was fired placing it in an appropriate orbit for a rendezvous and docking with the command-service module in the lunar-parking orbit. After the linkage of the modules and the performance of station-keeping duties, the two astronauts rejoined Astronaut Young in the command-service module, the hatch was secured and the ascent section of the lunar-excursion module was jettisoned. The command-service module continued to orbit the moon until the next morning, when the engine of the service module was reignited inserting the spacecraft into an earth trajectory. The service module was jettisoned just before reentry of the command module into the earth's atmosphere, and the command module, freed of all appendages, was turned so that its blunt end with the heat shield would be first to strike the earth's atmosphere. This procedure would protect the module from the intense heat generated by the impact of the space capsule traveling in excess of 24,000 miles an hour with the earth's atmosphere.

The Apollo 10 mission ended on May 26 with the splashdown of the spacecraft in the South Pacific, after eight days in space. During that time, it covered about half a million miles and completed 31 lunar orbits, approaching as near as 9.4 miles to its surface. Live television pictures were transmitted from the cabin of the spacecraft while journeying to and from the moon, and were carried by national and worldwide networks. *These were the first live colored television pictures to be transmitted from a spacecraft in space.* There were four programs, totaling 72 minutes, showing views of the earth, the moon, and the astronauts carrying on various activities in the cabin of the spacecraft.

The *Apollo 11* mission was initiated when the space capsule, on top of a Saturn V carrier vehicle, was hurled into an earth-parking orbit at 9:32 a.m. (EDT) July 16, 1969, carrying with it Neil A. Armstrong,[16] a civilian in command of the mission, accompanied by Colonel Edwin E. Aldrin, Jr.,[17] and Colonel Michael Collins.[18] The command-service module was named *Columbia* and the lunar-excursion module the *Eagle*. Less than twelve minutes following lift-off, the spacecraft entered a 118-mile earth-parking orbit where it remained for two and a half hours before being inserted into a lunar trajectory at a speed of 24,200 miles an hour.

At 1:48 p.m. the third stage of the carrier rocket was separated from the command-service module. The command-service module was then pivoted and linked with the lunar-excursion module to which the third stage was still attached. After about an hour, the third stage was jettisoned and placed in a solar orbit. The command-service module joined with the lunar-excursion module was turned to place the lunar-excursion module in line with the movement toward the moon. At this time, the Apollo 11 spacecraft was more than fifty thousand miles from earth.

On July 17, at 10:33 a.m. the spacecraft had reached a point equidistance from the earth and the moon and was traveling at 3,600 miles an hour. When it reached 148,000 miles from the earth, the astronauts began their first regularly scheduled colored telecast which continued 30 minutes. During this time, the camera photographed the earth and the astronauts rotated it to give the earth a rolling effect. The camera was then focused so as to take pictures of the inside of the cabin and the astronauts explained the various instruments and their functions. Astronaut Collins tossed a lighted flashlight into the air to demonstrate weightlessness.

The astronauts were awakened at 9:30 a.m. on July 18, at which time Apollo 11 was 73,000 miles from the moon, traveling at 2,400 miles an hour and rolling slowly to prevent overheating by the unfiltered rays of the sun. After finishing housekeeping tasks, Astronauts Armstrong and Aldrin began clearing the tunnel leading to the lunar-excursion module to permit Astronaut Aldrin to enter and inspect the lunar-excursion module. Their activities were televised and transmitted by Astronaut Collins in an unscheduled telecast beginning at 4:44 p.m. and lasting for 1 hour and 36 minutes. During this period, the spacecraft moved 2,300 miles closer to the moon.

Apollo 11 was only a few hours from the point where the earth's gravitational pull was equalized by that of the moon. The spacecraft entered the sphere of the lunar gravity at 11:32 p.m., when it was 214,963 miles from the earth and, as it drew nearer to the moon, its speed was accelerated.

On July 19 at 1:22 p.m., Apollo 11 entered a lunar orbit at which time it had traveled 244,930 miles from the earth. The astronauts spent most of the day beaming televised pictures of the moon to earth. At 6:58 p.m., Astronaut Aldrin crawled through the tunnel into the lunar-excursion module and powered it up for the first time since departing from earth.

On the morning of July 20, Astronauts Aldrin and Armstrong entered the lunar-excursion module in preparation for their moon-landing venture. At 1:47 p.m., while behind the moon and out of radio contact with Mission Control, the lunar-excursion module, the *Eagle*, was detached from the command-service module, the *Columbia*, and as it emerged from the far side of the moon, Astronaut Armstrong announced to an anxious world: "The Eagle has wings." The descent-stage engine was fired and the lunar-excursion module began its glide to the lunar surface. When it was 9.8 miles from the surface, the braking engine was activated and for about twelve minutes it descended and landed in a cloud of moon dust.

Commander Neil Armstrong announced at 4:17 p.m., EDT on July 20, 1969, to millions throughout the world: "Tranquility Base here: The Eagle has landed." Six and a half hours after landing, the astronauts emerged from their cabin, the intervening time since landing on the moon having been spent checking the control systems, sending back information to earth, and donning their spacesuits and back-packs. A television camera had been positioned to take pictures of man's first step on the lunar surface and to transmit it to viewers on earth.

The first to emerge from the cabin of the lunar-excursion module was Neil Armstrong who, after descending the ladder, paused on the pad of the leg of the lunar-excursion module before making man's first imprint upon the moon's surface at 10:56 p.m. As he made this imprint, he uttered the historic remark: "That's one small step for a man, one giant leap for mankind." Shortly thereafter, Astronaut Aldrin emerged from the cabin, descended the ladder and took his first step on the moon at 11:14 p.m.

The astronauts discovered that their movements were not greatly impeded and proceeded to carry out various extravehicular activities. They set up a television camera, a short distance from the lunar excursion module, to televise and transmit

Astronauts Armstrong and Aldrin Pose with the American Flag on the Moon (Source: NASA - Photo No. 69-H-1256)

Astronaut Aldrin Deploying the Passive Seismic Experimenter Package. The lunar ranging retro-reflector and the United States flag are already in place near the lunar excursion module. (Source: NASA - Photo No. 69-H-1254)

their lunar activities. These consisted of collecting samples of the moon's soil and rocks, placing the American flag in position on the surface of the moon, performing exercises to test their reactions under lunar conditions, and preparing for experiments by putting in place a seismograph, and a polished mirror for laser-beam experiments. They took time out to receive a message from President Nixon who spoke to them via radiotelephone from the White House.

Prior to lift-off from the moon, they had to package samples of the soil and rocks, seal the packages, and prepare the equipment, to be taken back to earth, for stowage in the cabin. Astronaut Aldrin ascended and entered the lunar-excursion module first in order to receive the various items as they were conveyed up to him. Upon completion of the loading activities, Astronaut Armstrong ascended the ladder, entered the cabin, and closed the hatch. The cabin was then pressurized, station-keeping duties were performed, back-packs and spacesuits were removed, and the former jettisoned. After these activities, the astronauts took their rest period before blasting off the lunar surface. The Apollo 11 astronauts extravehicular activities lasted for 2 hours and 21 minutes, during which time they collected 50 pounds of lunar soil and rocks to be taken back to earth.

At 1:55 p.m. on July 21, the *Eagle*, lifted off the moon in a blazing and flawless manner, after having been on the moon for 21 hours and 38 minutes. If the ascent engine had failed, the astronauts would have been condemned to death, because there were no back-up facilities nor could rescue operations have been effected before their power source failed and their supply of oxygen gave out. The *Eagle* rose rapidly and steadily and at 5:35 p.m., docked with the *Columbia* in a lunar-parking orbit 69 miles above the moon, piloted by Astronaut Collins. After a successful docking the two astronauts performed station-keeping duties, transferred the samples and other items to the command-service module, and then joined Astronaut Collins. The hatch to the ascent section of the lunar-excursion module was secured and shortly thereafter the lunar-excursion module was jettisoned.

They left behind them on the moon may items, including medals commemorating the American and Soviet spacemen who had lost their lives while engaged in space activities, a plaque bearing the astronauts' names and that of President Nixon, the American flag, a laser reflector, seismic detector, heat-absorbing panels, a television camera on a tripod, two portable life-support systems, two pairs of lunar boots, sample gathering tools, the descent stage of the lunar-excursion module and a silicon disk containing messages of goodwill from 73 heads of states. The weight of these items ranged from 4,100 pounds for the descent stage to less than one ounce for food bags and the total earth value of this moon junk was estimated at about one million dollars.

The Commemorative Plaque Left on the Moon Attached to the Descent Stage of the Apollo 11 Lunar Excursion Module. The plaque bears the signatures of President Nixon and the astronauts with the inscription: "Here men from the planet Earth first set foot upon the moon, July, 1969 A.D. We came in peace for all mankind." (Source: NASA - Photo No. 69-H-1261)

The critical maneuver, remaining to be executed, was to insert the command-service module into an earth trajectory. This was accomplished by the refiring of the service module's engine while the spacecraft was above the far side of the moon, during which time the spacecraft was out of contact with ground control. It entered the earth trajectory at 12:56 p.m. on July 22, and the astronauts were earthward bound. On July 23, at 7:03 p.m., they transmitted their final telecast during which they reflected on their moon-landing venture.

This historic mission was terminated on July 24 at 12:50 p.m., when the Apollo 11 spacecraft splashed upside down in the Pacific Ocean about eleven miles from the planned landing site, 250 miles south of Johnston Island and 950 miles southwest of Hawaii. It was quickly righted by its flotation balloons and shortly thereafter, the astronauts and the spacecraft were taken onboard the USS *Hornet*, where the astronauts were greeted by President Nixon. Even the President of the United States was not permitted to make physical contact with them, because of the rigorously enforced quarantine measures. The President had to converse with them by a microphone and gestures through a plate glass window of the isolation van in which they were to be confined for eight days. These precautions were deemed necessary to safeguard against contagion from some unknown microorganisms to which they might have been exposed during their stay on the moon.

REFERENCES

1. Transmitted to the Ground Control Station on landing on the moon.
2. The basic information for this chapter was derived from the *Semiannual Reports of the National Aeronautics and Space Administration,* leading newspaper releases, including *The New York Times* and *Aviation Week and Space Technology.*
3. *Mercury, Liberty Bell 7, Flight,* July 21, 1961 (p. 79); *Gemini III Flight,* March 23, 1965 (p. 83).
4. *Gemini IV Flight,* June 3, 1965 (p. 83).
5. *Report of Apollo 204 Review Board to the Administrator,* National Aeronautics and Space Administration, April, 1967.
6. After the tragic fire on January 27, 1967, the Apollo 4 was the first unmanned flight.
7. *Mercury, Sigma 7, Flight,* October 3, 1962 (p. 81); *Gemini VI-A Flight,* December 15, 1965 (p. 85).
8. *Gemini VII Flight,* December 4, 1965 (p. 84).
9. *Ibid.*
10. *Gemini VII Flight,* December 4, 1965 (p. 84); *Gemini XII Flight,* November 11, 1966 (p. 88).
11. *Gemini IV Flight,* June 3, 1965 (p. 83).
12. *Gemini VIII Flight,* March 16, 1966 (p. 85).
13. *Gemini VI-A Flight,* December 15, 1965 (p. 85); *Gemini IX-A Flight,* June 3, 1966 (p. 86).
14. *Gemini III Flight,* March 23, 1965 (p. 83); *Gemini X Flight,* July 18, 1966 (p. 86).
15. *Gemini IX-A Flight,* June 3, 1966 (p. 86).
16. *Gemini VIII Flight,* March 16, 1966 (p. 85).
17. *Gemini XII Flight,* November 11, 1966 (p. 88).
18. *Gemini X Flight,* July 18, 1966 (p. 86).

Chapter IX*

INITIAL ON-THE-SURFACE LUNAR EXPLORATIONS OF THE UNITED STATES (Apollo Missions 12, 13 and 14)

The Apollo Program, as originally planned, consisted of 20 lunar missions, but following the successful landing of men on the moon by July 1969, it was revised, because of budget limitations. In January 1970, it was announced that the eighth lunar-landing mission was cancelled and that the remaining seven missions were to be scheduled over a period extending to 1974. Four missions were to take place prior to the end of 1971 and the remaining three after the completion of the space station program (Skylab) to be initiated in 1972.[1] A further retrenchment was announced in September 1970. Two additional lunar landing missions were cancelled, meaning that the last three missions would not be undertaken. Two of the four remaining missions were to be scheduled for 1971 and two for 1972.

The manned Apollo 12 spacecraft was launched on a planned ten-day orbit from the Kennedy Space Center on November 14, 1969, at 11:12 a.m. having as its destination a lunar site near Landsberg Crater on the Ocean of Storms, about nine hundred and fifty miles west of where Apollo 11 had landed on the Sea of Tranquility. A leaky tank threatened to delay the blast-off, but the replacement of the tank was achieved in time to allow the spacecraft to lift off on schedule. The weather was very unfavorable and the spacecraft was struck twice by lightning almost immediately after lift-off knocking out temporarily its power system. Fortunately the battery-operated secondary system responded, and the spacecraft continued its ascent and assumed an earth orbit at an altitude of 118 miles within 11 minutes and 40 seconds after lift-off. An onboard clock appeared to be the only thing damaged by the lightning strike.

The mission was under the Command of Commander Charles Conrad, Jr., a veteran of two space missions,[2] who was accompanied by Lieutenant Commander Richard F. Gordon, Jr., pilot of the command-service module, the *Yankee Clipper,* also a veteran of two previous missions,[3] and Commander Alan L. Bean, pilot of the lunar-excursion module, the *Intrepid,* making his first journey into space.

*This chapter covers those Apollo missions where the lunar exploratory activities of the astronauts were confined to the distances they could cover on foot.

The primary purposes of the Apollo 12 mission were to develop techniques for achieving pinpoint navigation and landings on the moon, deploy a nuclear-powered scientific station and an Apollo Lunar Surface Experiments Package (ALSEP), perform geological inspections, survey an extended area of the surface of the moon, determine man's capabilities for working under the lunar conditions, and photograph potential landing sites for future Apollo missions. In addition, the astronauts were to walk to the site of Surveyor 3 and retrieve certain parts.[4]

The spacecraft was inserted into a translunar hybrid trajectory[5] by a midcourse correction which would bring it closer to the moon, save fuel and permit a moon landing with good visibility. On previous Apollo missions free return trajectories had been used largely for safety reasons. With the free return trajectory, if the spacecraft should fail to enter a lunar orbit, it would swing around the moon and be flung into an earth trajectory without firing of the spacecraft's engines. With the hybrid trajectory, the spacecraft would approach too close to the moon for an accurate return to the earth without the use of spacecraft's engines. In this particular instance, if the spacecraft should fail to enter a lunar trajectory it would miss the earth by some 56,000 miles unless some midcourse corrections were made, which could probably be accomplished by utilizing the descent-rocket engine of the attached lunar-excursion module. The hybrid trajectory would bring the spacecraft within about seventy-four miles of the moon at its closest approach, whereas using the free trajectory, it would bring it within eighteen hundred miles.

The television camera transmitted from inside the cabin during the outward voyage as the astronauts checked the equipment and prepared for the rocket firing for a midcourse correction of the trajectory. The correction was made to permit the lunar-excursion module to land as the sun rose over the Ocean of Storms and when the astronauts were in contact with the primary ground station at Goldstone, California.

On November 17, Apollo 12 assumed a lunar orbit at 10:57 p.m., after a six-minute rocket firing slowing the spacecraft from 5,574 to 3,600 miles an hour and allowing it to be captured by lunar gravity. The initial orbit was elliptical with an apolune of 194 miles and a perilune of 72 miles, but it was later made more circular with an apolune of 70 miles and a perilune of 62 miles. On the first orbit, the astronauts initiated a color telecast showing the surface of the moon, focusing primarily on the white central peaks of the Crater Langrenus.

After orbiting the moon ten times, Astronaut Conrad and Bean entered the lunar-excursion module on November 18, shortly before 6:00 p.m. (EST) to prepare for the descent to the surface of the moon. At 11:16 p.m. after completing two more orbits, the lunar-excursion module was separated from the command-service module, the *Yankee Clipper*, and 30 minutes later, Commander Gordon moved the command-service module a safe distance away from the lunar-excursion module, by firing its engine.

The *Intrepid* made a successful landing on the lunar target at 1:54 a.m. on November 19. The landing site was on the eastern side of the Ocean of Storms within 1,120 feet of the place on the Sea of Tranquility where Surveyor 3 had landed in 1967. The precise landing was possible because of the skill of the astronauts who had been trained to execute pinpoint navigation and of changes in the descent and landing procedures implemented to prevent overshooting the target as was the case with the Apollo 11 landing.[6] In this instance, Commanders Conrad and Bean had to land within 3,000 feet of the site of Surveyor 3, if they were to walk to it and retrieve some of its parts.

The first extravehicular activities were undertaken at 6:44 a.m. on the day they landed and lasted until 10:31 a.m. Astronaut Conrad was first to step upon the moon's surface and was followed within 30 minutes by Astronaut Bean. All of their extravehicular activities were not televised because the camera was damaged shortly after it was removed from the lunar-excursion module by being exposed inadvertently to the intense sunlight. They collected contingency samples of the lunar soil and rocks, deployed the S-Band antenna, placed in position the Solar Wind Composition Experiment, and erected the American flag. In addition, they unloaded the Apollo Lunar Surface Experiments Package (ALSEP) and deployed it about six hundred feet from the lunar-excursion module. It consisted of a passive seismometer for measuring the seismic activity of the moon, a magnetometer for measuring the strength, velocity and direction of the electrons and protons radiating from the sun and striking the lunar surface, a suprathermal-ion detector for measuring the nature of positive ions close to the surface of the moon, a cold cathode-ion gauge, for ascertaining the density of any ambient lunar atmosphere, and a detector for measuring the accumulation of dust on the ALSEP. This array of experiments, more elaborate than that left by the Apollo 11 astronauts, was also to be left to transmit scientific and engineering data to earth for a year or longer. Later the astronauts returned to the lunar-excursion module, entered the cabin, closed the hatch, repressurized the cabin, and rested for five hours.

The second moon excursion began at 10:55 p.m., on the same day, and lasted 3 hours and 49 minutes, during which time the astronauts walked to the site of the ALSEP experiments, the rims of several small craters, the site of Surveyor 3, and back to the lunar-excursion module. They moved from 1,500 to 2,000 feet from the lunar-excursion module covering a total distance of about 6,000 feet taking photographs, collecting samples of the moon's soil and rocks, and retrieving parts from Surveyor 3, including a camera, and then returned to the module at 5:54 a.m. Before reentering the cabin of the lunar-excursion module for the last time, they recovered the Solar Wind Composition Experiment aluminum sheet.

One of the significant achievements of this mission was the introduction of atomic power on the surface of the moon, when on the morning of November 19, Commander Bean placed a radio-active fuel cell in the System for Nuclear Auxiliary Power (SNAP) 27 generator. This generator, with no moving parts, was to be used to generate electricity to energize the array of instruments to be left on the moon.[7] It could make electrical energy available on the moon for years to come.

Commander Richard F. Gordon, piloting the command-service module in the lunar-parking orbit, was engaged in conducting experiments and taking close-up pictures of the moon. He had been assigned to take 150 multi-spectra photographs of its surface, using various colored filters. This required that he keep the command-service module on a precise course and, at the same time, monitor the four cameras focused on the same spot on the lunar surface. Each camera took its pictures through a different colored filter: red, blue, green, and infrared, and the pictures were to be combined later into a composite picture indicating the locations of various minerals on the surface of the moon.

On November 20 at 9:26 a.m., after a stay of 31 hours and 31 minutes on the moon, the two astronauts, in the ascent section of the lunar-excursion module, the *Intrepid,* blasted off to rejoin Commander Gordon in the command-service module, the *Yankee Clipper.* The descent stage of the lunar-excursion module, which served as a launching platform, was left behind together with a nuclear power station, an

Close-up Lunar Soil, Apollo 12 Mission (Source: NASA - Photo No. 69-HC-1348)

Close-up View of Lunar Rocks, Apollo 12 Mission (Source: NASA - Photo No. 69-HC-1346)

array of instruments, the American flag and a plaque bearing their names and that of the President. No difficulties were encountered in accomplishing a rendezvous and docking with the mothership at 12:58 p.m. after which they transferred the cargo and themselves to the command-service module. The hatch was then secured and the ascent stage of the lunar-excursion module was jettisoned and placed in a lunar trajectory. It was anticipated that it would impact the moon about forty-five miles from the site of the seismometer at about 5,000 miles an hour.[8]

The *Yankee Clipper,* with the service-module attached, continued in its lunar orbit to enable the three astronauts to complete their photographic assignment, giving special attention to possible landing sites. It left the lunar orbit on November 21, after 45 revolutions of the moon, and assumed an earthward trajectory altered on November 22 when the spacecraft was 208,000 miles from the earth. On November 23, from 7:30 to 8:06 p.m., *the astronauts held the first space-to-ground news conference.* While the spacecraft was 108,000 miles away in space and traveling at 3,670 miles an hour, the astronauts responded to questions that had been submitted previously in writing to the Manned Space Station and transmitted to them.

The Apollo 12 mission was terminated on November 24, at 3:58 p.m., when the Apollo capsule, containing the astronauts, achieved a pinpoint-landing on a stormy Pacific Ocean, 404 miles southeast of Pagopago and 2,651 miles south-southwest of Honolulu. The capsule splashed down three miles from the recovery ship, USS *Hornet,* bottom-up and was soon righted by its flotation balloons. Shortly thereafter they were hoisted aboard a rescue helicopter and flown to the aircraft carrier, where they were warmly greeted and received a congratulatory telephone message from President Nixon, informing them that they had been promoted to the rank of Captain and reminding them of his invitation for them to dine with him at the White House.

The astronauts brought back with them two boxes of lunar soil and rocks, weighing about forty-five pounds each, together with a bag containing about fifty additional pounds. In addition, they returned with a television camera, a painted aluminum strut, an electrical cable[9] and a robot-like mechanical scoop used for digging in the lunar soil, all recovered from Surveyor 3.

The Apollo 13, ten-day lunar landing mission was plagued with mishaps threatening to postpone or abort the mission before and after the launching. The mission ultimately ended in failure midway between the earth and the moon. The final stroke of misfortune introduced a tension-filled period, during which it was uncertain whether the crippled spacecraft and its crew would make it back to earth safely.

During the preparatory stages of the mission, the engineers confronted a helium pressure tank problem which did not prove to be serious enough to delay the scheduled lift-off. Further concern arose because a member of the crew, Lieutenant Commander Thomas K. Mattingly II, pilot of the command module, the *Odyssey,* had been exposed to German measles and was replaced two days before the scheduled launching. Captain James A. Lovell, Jr.,[10] commander of the mission, and Mr. Fred W. Haise, Jr., a civilian astronaut and pilot of the lunar-excursion module, the *Aquarius,* were permitted to remain as members of the crew, following thorough medical examinations. Mr. John L. Swigert, Jr., a civilian back-up pilot for the mission, was selected to replace Lieutenant Commander Mattingly. The possibility of a strike at a key ground control station in Australia also threatened postponement of the mission. Fortunately, the strike issues were compromised in time so as not to interfere with the scheduled lift-off.

Surveyor 3 Spacecraft on the Surface of the Moon Being Examined by One of the Astronauts of the Apollo 12 Mission, with Lunar Excursion Module in the Background (Source: NASA - Photo No. 69-H-1986)

Finally, Apollo 13 flight was cleared for launching a day before the scheduled time for take-off on a mission which proved to be the most hazardous of the American manned space program. If the lift-off had not taken place when it did, it would have had to be postponed for a month until Lieutenant Commander Mattingly had passed the incubation period and when the light conditions at the lunar-landing site were favorable. The launch took place at 2:13 p.m. (EST) on April 11, 1970, and following a smooth countdown, the spacecraft rose from its pad and arched over the Atlantic Ocean. The booster-stage engines cut off on schedule after which the five engines of the second stage ignited, but the center engine shut down 2 minutes and 7 seconds ahead of time. The spacecraft continued to be boosted upward despite the premature shut down by longer burnings of the four sustaining engines, prior to the separation of the second stage. Within eleven minutes after take-off, the command-service module, with the third stage attached, entered an earth orbit, with an altitude of 117 miles.

The third-stage rocket fired at 4:48 p.m., after the spacecraft had circled the earth for two and a half hours, placing it in a lunar trajectory. Later, the third stage was jettisoned and placed in a lunar-crash trajectory. Its impact with the surface of the moon would create a man-made moonquake, and would be recorded by the seismographs left on the moon by the astronauts of the Apollo 11 and Apollo 12 missions. The command-service module linked to the lunar-excursion module continued on its course to the moon.

The Lift-off of the Apollo 13 Spacecraft (Source: NASA - Photo No. 70-H-487)

The selected landing site for Apollo 13 was Fra Mauro, a formation of lunar debris, believed to have been pushed outward when the basin of the Sea of Rains was formed by the impact of an immense meteorite. This region was selected on the assumption that samples of the lunar soil and rocks could be collected which had not been materially altered for four billion years. The landing site was a relatively level area between two clusters of craters, Doublet and Triplet.

The astronauts were to spend approximately thirty-four hours on the moon and make two extravehicular excursions each lasting about four hours. During the first moon-walk, they were to spend the greater part of the time drilling three holes into the moon's surface and setting-up an atomic battery-power station as a source of energy for the series of experiments making up the ALSEP. In addition, they were to walk to the Star Crater located less than a mile from the planned landing site.

On the second excursion, much of the time was to be used for collecting samples of lunar soil and rocks, each of which was to be photographed before and after it was collected and then bagged. The astronauts were to walk also to the rim of Cone Crater, rising some four to six hundred feet above the surrounding terrain thought to have been blasted out of the top of a long sloping range similar to the Fra Mauro formation. This overall excursion was to cover in total 8,700 feet over an area changing from fine dusty soil to rugged terrain as it sloped upward to the crater's rim, and would require about four hours to complete.

All systems of the Apollo 13 spacecraft were functioning normally as it coasted toward the moon on a trajectory which, with a mid-course correction, would place it in an orbit with a minimum height of 69 miles above the moon. The spacecraft passed the midpoint between the earth and the moon on April 12, at 5:34 p.m., traveling at 3,400 miles an hour, and everything pointed to a third successful landing on the moon.

At 10:15 p.m. on April 13, when the spacecraft was approximately two hundred thousand miles from earth, the astronauts informed Mission Control at Houston that a warning light indicated an electrical power problem. Captain Lovell reported low oxygen readings on two of the three power-producing cells and within five minutes, it was discovered that the spacecraft was venting, what looked like gas, causing it to pitch and roll. Within a half an hour, fuel cell three was out, the supply of oxygen in the cabin had dropped, and one of the two main electrical circuits was dead. Three systems were flashing warning lights by 10:55 p.m.,[11] and by 11:20 p.m., the venting was causing a positive pitch which could only be stopped by thruster jet firings.

The Moon as Seen from the Apollo 13 Spacecraft during Homeward Journey. Shown in the photograph are the Sea of Crisis, Sea of Fertility, Sea of Tranquility, Sea of Serenity, Sea of Nectar, Sea of Vapors, Border Sea, Smyth's Sea, Crater Langrenes, and Crater Tsiolkovsky. (Source: NASA - Photo No. 70-H-689)

Mission Control instructed the astronauts to activate an emergency battery and informed them that it was considering implementing the life-boat procedure.[12] By 11:59 p.m., all fuel cells in the command-service module were off and there remained only 15 minutes of electrical power. The astronauts were instructed to move into the lunar-excursion module, leave the hatches open, and activate its systems. Astronaut Swigert remained in the command module's cabin to turn off its systems, being supplied with oxygen from the lunar-excursion module through the tunnel and open hatches. The hatches were to remain open because there was insufficient room in the lunar-excursion module, designed to accommodate two not three astronauts. Two of them were to sleep in the command module while the other was to remain awake in the lunar-excursion module to monitor its systems.

The moon landing was cancelled at 12:13 a.m. April 14 and plans were instituted to return the Apollo capsule and the astronauts to earth safely, if possible. It was decided that the spacecraft should continue on its lunar course and swing around the moon, using the moon's gravity to bend its course toward the earth. The descent engine of the lunar-excursion module was to be used to refine the course. This procedure would take less fuel, provide the shortest return time, and was considered the safest for getting the astronauts back.

There developed also a carbon dioxide problem, caused by the failure of command module cabin's lithium hydroxide filter system overloading the lunar-excursion module's filter system. In addition, the water for cooling the electronic equipment was so low that it had to be conserved for carrying out the midcourse correction necessary to bring the spacecraft into the narrowly defined reentry angle, no greater than 7.3 and no less than 5.8 degrees. It was feared that the spacecraft would have to coast to earth without the inertial-guidance system because of inadequate water to keep it cool. In fact, a day before the splashdown, the astronauts ran out of water and had to do without it for the remainder of the return trip. On April 15, the rocket engine of the lunar-excursion module was fired at 9:41 p.m. and the spacecraft pulled away about six thousand miles from the right side of the moon, placing it in an earth trajectory.

A successful midcourse burn of 15 seconds, was accomplished at 11:32 p.m. on April 15, eleven minutes earlier than planned, when the spacecraft was 175,000 miles from the earth. If the burn had failed the spacecraft would have missed the earth by 104 miles but the astronauts would still have had the small control jets on either side of the lunar-excursion module and the command-service module to fall back on.

Before the midcourse burn, concern arose because of a flashing battery alarm indicating that one of the six batteries in the lunar-excursion module was overheating. Fortunately, this turned out to be a false alarm. In addition, to these problems, the astronauts had to endure the low temperatures, which hindered the performance of their activities and impaired their rest. In preparing for reentry, they had to work under conditions subject to 38 degrees Fahrenheit, while clad in coveralls. As they neared the end of their journey, their activities became intensified, because of the need for going over checklists with Mission Control for a period of more than four hours. Added to all this, the batteries in the cabin of the command module had to be recharged from those of the lunar-excursion module, for executing the final stages of reentry and return to earth.

The astronauts moved back into the cabin of the command module about six hours before the scheduled splashdown. The crippled service module was jettisoned at 8:23 a.m. and photographs were made of it tumbling in space. The separation was accomplished by the push-pull technique, necessitating firing the lunar-excursion module's engine to push the service module away from the command module and at the same time setting off the device in the command module to effect the separation. After this maneuver, the direction of the lunar-excursion module, with the command module attached, was reversed in order to avoid colliding with the service module.

The lunar-excursion module, the *Aquarius*, which had proved so trustworthy, was jettisoned at 11:43 a.m., the command module powered up and reentry started at 12:53 p.m. The spacecraft splashdown in the Pacific Ocean, on April 17 at 1:08 p.m. within five miles of the recovery ship, *USS Iwo Jima* and 800 yards from the pre-set landing point located 610 miles southeast of the American Samoa. The first helicopter was overhead within three minutes of splashdown and the astronauts were aboard within thirty minutes and on their way to the recovery ship. They did not have to undergo an extensive isolation period because they had not landed on the moon. The Apollo 13 Flight had lasted 5 days, 22 hours, and 53 minutes and covered more than half a million miles.

At the end of this hazardous journey, there remained enough water for 21 hours, oxygen for 125 hours, electricity for 31 hours and lithium-hydroxide for 38 hours. The astronauts, as a precaution, had drained six plastic bags of water from the lunar-excursion module and placed them in the command module in case of need before and after landing.

During those crucial hours in the crippled spacecraft, the World was stirred to sympathy and a desire to assist in whatever manner possible. Twelve nations offered to aid in the recovery operation, if needed, and numerous messages of concern were received and prayers were offered for their safe return. The United States Senate and House of Representatives passed resolutions on April 14, urging all Americans at 9:00 p.m. that night to pray for their safe return and that special religious services be held for them. President Nixon flew 5,000 miles from Washington to Honolulu to honor them and on April 18, bestowed on each the Medal of Freedom, the Nation's highest non-military honor.

One experiment was successfully accomplished in spite of the cancellation of the moon landing. The spent third-stage rocket crashed with a force equal to 11 tons of TNT on April 14, at 8:09 p.m., on the Ocean of Storms, 87 miles west-northwest of the landing site of Apollo 12. The 15-ton rocket caused tremors which lasted for 3 hours and 20 minutes and were recorded on the seismographs left on the moon by the two previous Apollo missions. The mushroom cloud of moon-dust, caused by the impact, spread outward from the point of impact at more than 14,000 miles an hour.

On January 31, 1971[13] the Apollo 14 mission was initiated, under the command of Captain Alan B. Shepard, the first American to fly in space,[14] accompanied by Commander Edgar D. Mitchell, pilot of the lunar-excursion module, the *Antares*, and Major Stuart A. Roosa pilot of the command-service module, the *Kitty Hawk*. A scratch was found on the outside of the second-stage rocket as the tower was being removed, but it proved to be so superficial that it did not endanger the performance of that stage. The lift-off was delayed for 40 minutes and 2 seconds because of a threatening electrical storm,[15] but at 4:03 p.m. it blasted off and in about twelve minutes the spacecraft assumed a low earth orbit 117 miles high from which it was inserted into a lunar trajectory at 6:31 p.m. after orbiting the earth for two and a half hours.

The mission's primary goal was scientific exploration of the moon. The two previous Apollo missions had as their primary objectives testing the capabilities of

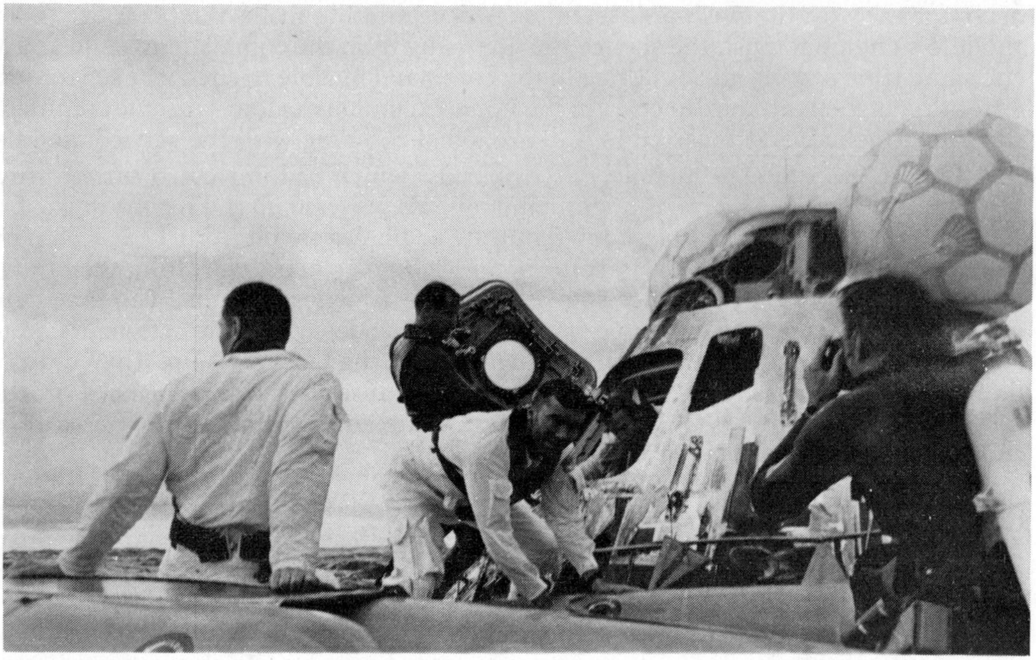

The End of the Unsuccessful Lunar Flight of Apollo 13 — Recovery of the Astronauts and the Space Capsule in the South Pacific (Source: NASA - Photo No. 70-H-644)

the equipment and the procedures for the landing of men on the moon and their safe return to earth. The objectives of this mission included installing an atomic-battery power station, deploying a package of highly sophisticated scientific instruments (ALSEP), and setting up marker flags to align geophones for detecting the vibrations of small explosive charges at varying distances from seismographs. The astronauts were also to align instruments to continue operating after they had departed from the moon and to leave on the descent stage of the lunar-excursion module a station to transmit data recorded by these instruments.

Several precautionary measures were taken to avoid the situations encountered on the Apollo 13 mission. A preflight isolation period was inaugurated, termed Flight Crew Health Stabilization, to protect the astronauts and back-up astronauts, as well as their wives and personnel directly involved in the flight from common viruses which might cause the postponement of the mission. Modifications were also made in the spacecraft. An extra oxygen tank was installed and isolated from the other oxygen tanks in the service module. The tanks were rewired, with the wires encased in metal conduits and some were coated with "teflon" to provide additional protection. An additional battery was installed as a source of emergency power and fans and switches were removed from the tanks. In addition, a tank was installed to safeguard against shortage of water.

In spite of these measures, the Apollo 14 crew encountered certain difficulties while in flight. At 7:11 p.m., about three hours after lift-off, Major Roosa attempted to effect a soft-docking between the lunar-excursion module, to which the third-stage rocket was attached, and the command-service module. This maneuver had never failed on previous missions, but he had to make five attempts before he was successful. On the sixth attempt at 9:00 p.m., when the spacecraft was 23,800 miles from earth, using an unorthodox docking procedure, he achieved a firm linkage of the modules. At 9:32 p.m. the third-stage rocket was jettisoned and put into a separate lunar trajectory causing it to crash on the surface of the moon with an estimated impact of eleven tons of TNT. This man-made moonquake was recorded by the two seismographs stationed on the moon.[16] A similar impact occurred when the lunar-excursion module's ascent stage was jettisoned and crashed on the surface of the moon, traveling at 3,754 miles an hour.

The two hours required for docking caused considerable discussion relating to the advisability of proceeding with a moon landing and it was not until February 4, just before the spacecraft assumed a lunar orbit, that the decision to go for a landing was finalized. On February 1, the astronauts fired the main rocket of the spacecraft for ten seconds to achieve a better aim at the moon, because without the midcourse correction, the spacecraft would have missed it by 2,419 miles. The correction in the trajectory would bring it within seventy miles of the moon and make up most of the 40 minutes lost because of the delay in lift-off. The adjusted aim proved to be seven miles too far away and shortly before entering the lunar orbit, another correction was made in the trajectory.

The Apollo 14 spacecraft entered the lunar orbit at 2:00 a.m. at a speed allowing it to be captured by the moon's gravity. A second firing, after two orbits of the moon, was to lower the orbit to 66 miles on the far side of the moon and ten miles on the near side. One of the batteries in the ascent stage of the lunar-excursion module showed 36.7 or 3/10 of a volt degradation, but this was not considered serious enough to abort the landing.

The *Antares* and the *Kitty Hawk* separated at 11:51 p.m. on February 5, during the twelfth revolution and the *Kitty Hawk* moved into a lunar-parking orbit with an apolune of 69 miles and a perilune of ten miles. During the final descent of the *Antares,* an abort signal caused concern and immediate action was taken to rewrite the program clearing the way for the landing.[17] On February 5, at 4:18 a.m., the *Antares* landed on a relatively level valley in the Fra Mauro highlands, a cratered and rock covered area, 110 miles east of where the Apollo 12 had landed in November 1969.

Captain Shepard opened the hatch at 9:49 a.m. and stood on the platform at the top of the ladder and then descended to the moon's surface at 9:57 a.m. He was followed in about five minutes by Commander Mitchell. Their first extravehicular excursion, of about 1,000 feet, lasted for 4 hours and 40 minutes, and involved collecting contingency soil and rock samples, deploying a television antenna on a tripod and an umbrella-like antenna, and erecting the American flag. They set up a nuclear scientific station, similar to that established by Apollo 12 astronauts to record moonquakes, measure electrically charged particles, and detect the composition and energy of the solar wind. They also exploded charges which set in motion vibrations at a depth of 70 feet for determining the nature of the moon's subsoil. In the midst of these activities, they received an invitation from the President, relayed to them by Mission Control, to dine at the White House upon their return to earth.

The astronauts set out on their second moon excursion early on the morning of February 6, following a rest period of 4 hours and 39 minutes, realizing that it might be shortened because of a small leak in Commander Mitchell's spacesuit. They walked toward the rim of a crater about a mile from the lunar-excursion module, pulling a two-wheel cart. Although they did not reach the rim of the Cone Crater, the mission was, scientifically speaking, successful.

Astronauts Shepard and Mitchell spent over nine hours outside the lunar-excursion module of which 7 hours and 35 minutes were spent in moon-walking. *During these activities, Captain Shepard introduced golf on the moon.* He made two drives using a club which he had fashioned by attaching an iron to the handle of one of the tools. The first drive he made carried the ball about two hundred yards and the second about four hundred. If the drives had been made on earth, the balls would have gone probably no more than one hundred and forty yards.

The ascent stage of the *Antares,* blasted off the moon at 1:49 p.m., on February 6, after 33½ hours on the moon's surface. The astronauts took with them about ninety-six pounds of lunar soil and rock[18] *the largest and most varied sample, probably containing some of the oldest rocks ever removed from the moon.* They left behind the American flag, a series of scientific instruments, a nuclear scientific power station, an antenna, the descent stage of the lunar-excursion module, and other discarded objects.

At 3:35 p.m., about two hours after the lunar lift-off, the ascent stage of the lunar-excursion module rendezvoused and docked with the command-service module without difficulty. The ascent stage was jettisoned at 5:58 p.m. and crashed later on the moon south of the landing site of Apollo 12. The engine of the command-service module was fired when it was behind the moon, putting it in an earth trajectory and again at 1:37 p.m. on February 7 when the astronauts fired a short burst of the maneuvering rockets to change the course 7/10 of a foot.

On the homeward journey, they carried out a series of scientific experiments, never before conducted, under the conditions of zero gravity while in space. The purposes of the experiments were to determine the effect of zero gravity on welding, liquid and heat flows, together with the separation of molecules in a solution. In addition, a space-to-earth news conference was held for about twenty-five minutes.

The Apollo 14 space capsule splashed down on target in the Pacific Ocean at 4:05 p.m., on February 9, about nine hundred miles south of the American Samoa. The astronauts were aboard the recovery ship USS *Orleans* within an hour and immediately went into a three-week quarantine. While on board the carrier, they received a telephone message from President Nixon congratulating them on their achievements and welcoming them back to earth.

REFERENCES

1. The Skylab Program was initiated on May 14, 1973, when the first space station of the United States was sent aloft.
2. Commander Charles Conrad, Jr., participated in *Gemini V Flight* August 21, 1965 (p. 84) and *Gemini XI Flight,* September 12, 1966 (p. 88).
3. Lieutenant Richard F. Gordon participated in *Gemini XI Flight,* September 12, 1966 (p. 88).
4. *Surveyor 3 Mission,* April 17, 1967 (p. 50).
5. A compromise between two different paths or trajectories which could be taken to the moon.
6. Apollo 11 overshot its planned landing target by 22,380 feet or about four and a half miles.

7. The array of instruments left on the moon by the Apollo 11 astronauts relied on solar-panel batteries for their energy, which operated only during the lunar-day, lasting two weeks.
8. The crash of the lunar-excursion module caused a man-made moonquake, the vibrations of which were recorded for 55 minutes by the seismometer left on the moon by the previous Apollo astronauts.
9. The electrical cable had been wrapped in aluminum foil which covered a known number of microorganisms. The recovery of the cable should permit the scientists to determine, if any, the number of microorganisms which survived the lunar-environmental conditions.
10. Captain James A. Lovell, Jr., participated in *Gemini XII Flight,* November 11, 1966 (p. 88); and *Apollo 8 Flight,* December 21, 1968 (p. 96).
11. At this time, it was apparent that there was little hope for completing the moon landing and the major problem was how to return the astronauts to earth safely.
12. The astronauts were to use the lunar-excursion module as a lifeboat to aid them in getting back to the point where they were to prepare for reentry into the earth's atmosphere.
13. This date marked the 13th anniversary of the first satellite launching of the United States, January 31, 1958, when a 70-foot Jupiter C rocket blasted the 30.5 pound Explorer I into an earth orbit (p. 43). It was also the 10th anniversary of the space flight of Ham, the chimpanzee, which took place on January 31, 1961, when a Resstone rocket, with 1/100 the power of a Saturn V rocket, sent the chimpanzee on a 15-minute suborbital flight (p. 78).
14. *MR - 3 Flight,* May 5, 1961 (p. 79).
15. It was ruled, following the striking of the Apollo 12 spacecraft by lightning after it took off under unfavorable conditions, that no launching would be undertaken unless weather conditions were considered favorable and there was no danger of an electrical storm.
16. The spent third stage of the Saturn V rocket of the Apollo 14 mission crashed on the moon on the Ocean of Storms, south of the Apollo 12 landing site about thirty minutes after the Apollo 14 spacecraft went into a lunar orbit.
17. The problem was apparently due to a faulty cockpit switch which produced intermittent short-circuits.
18. The combined weight of the samples of lunar soil and rocks brought back by the astronauts of Apollo 11 and Apollo 12 missions was about one hundred and twenty-four pounds.

*Chapter X**
MOTORIZED ON-THE-SURFACE LUNAR EXPLORATION ACTIVITIES OF THE UNITED STATES (Apollo Missions 15, 16 and 17)

The *Apollo 15* spacecraft blasted off from its pad on July 26, 1971 at 9:34 a.m. (EDT) for a 12-day roundtrip to the moon, after one of the smoothest countdowns of the Apollo missions.[1] The mission was most ambitious being highly scientific in its orientation. Colonel David R. Scott, a veteran of two previous space flights,[2] was in command; Lieutenant Colonel James R. Irwin, pilot of the lunar-excursion module, the *Falcon*, and Major Alfred M. Worden, pilot of the command-service module, the *Endeavor*. The major purposes of the mission were to study the origin of Hadley Rille, gather samples of "mascon," dense lunar material which exerts a slightly higher gravitational force than less dense material, and drill three holes in the lunar surface for measuring the flow of heat from the moon's interior.

The astronauts had been thoroughly trained as part of a large team, for more than a year and a half, to work together as a unit learning how to cope with situations which they would likely encounter on the moon in regions contiguous to Hadley Rille (Hadley-Apennine Site). An essential aspect of their training was to learn to work as a team, when part of it was on the moon and part was on earth. They studied the fundamentals of geology and the techniques of collecting and examining soils and rocks.

Considerable field work was conducted in the gorges of the Rio Grande River in New Mexico and the Little Colorado River in Arizona to acquire an understanding of the kinds of terrain they would be exploring in the regions where the lunar-excursion module was scheduled to land. Time was spent at the Meteorite Crater in Arizona and other crater sites in Nevada and Canada. In all fifteen locations were visited during their preparation for this lunar landing mission.

The landing site of the Apollo 15 lunar-excursion modules was located in a region made up of a wide variety of physical features, 465 miles north of the lunar

*This chapter covers those Apollo missions of which the exploratory activities of the astronauts were supplemented by the lunar rover, making it possible to cover more of the moon's surface than could be covered on foot.

equator on the southern edge of the Sea of Rains. It was about two miles south of the site where the Apennine Mountains rose some twelve thousand feet above Palus Putredinis (Putrid Swamp) that formed a curved wall surrounding the largest crater formation on the moon, the Sea of Rains (Mare Imbrium). Three miles to the north were three dome-like features believed to have been formed either by volcanic action or by the impact of a huge meteorite. Hadley Rille, a mile-wide canyon-like gorge, about twelve hundred feet deep and twisting across the terrain, was one of the most impressive features of this region.

Compromises were made as to the items making up the payload and their respective weights, because of the more sophisticated scientific instruments and the 460-pound four-wheel lunar rover operated by solar batteries being taken to the moon for the first time. The back-up camera was dispensed with and other items were restricted in order to make room for an additional 5,110 pounds of cargo. The combined weight of the command-service module and the lunar-excursion module was 103,100 pounds as compared with 97,990 pounds for the Apollo 14 mission. This was to be the heaviest space vehicle put into an earth obit up to that time. To achieve this, the five main rockets of the first stage of Saturn V were set to burn four-tenths of a second longer than on past Apollo missions, and four of the eight retrofiring rockets, used for the separation of the first stage from the second stage were eliminated. The lunar-excursion module was equipped also with larger propellant fuel tanks.

The moonward journey of Apollo 15 was uneventful and the nose-to-nose docking of the command-service module with the lunar-excursion module was accomplished without difficulty. Later, a faulty cockpit switch caused some concern, but it was not serious enough to halt the mission. A firing of the main rocket indicated that it would not hinder the moon landing. A water leak was discovered on July 29, after the regular injection of chlorine into the water supply, but was quickly repaired by using a wrench from the tool box. Additional worries arose when a change in the electrical voltage tripped a circuit breaker in the command cabin and caused several control lights to go out. Again this malfunction proved more annoying than serious. Furthermore, a broken glass covering on one of the instrument dials in the lunar-excursion module cabin caused some concern but not of a serious nature.

The first scientific experiment was conducted during the moonward flight and lasted for about an hour. The blindfolded astronauts observed flashes, which were thought to be connected with cosmic radiation. They saw about sixty of them, some in the form of streaks and others similar to the flashes made by flashbulbs exploding in the dark. This first test was carried out when the spacecraft was about 188,000 miles from the earth and later two more were made; one when Major Worden was alone in the command-service module in the lunar-parking orbit and the other when the astronauts were together in the cabin of the command module on the return journey to earth.

The spacecraft went into a lunar orbit on July 29 and circled the moon twice before firing its rocket to assume a lower orbit. The separation of the lunar-excursion module was delayed when a high temperature reading was shown on two electrical power wires loosely plugged into the docking mechanism, but the malfunction was corrected and the separation took place at 2:15 p.m. The descent path of the lunar-excursion module was steepened to 25 degrees, nine degrees steeper than for other landings, to ensure that the module would clear the Apennine Mountains. It touched down 1,500 feet northeast of the planned target at 6:16 p.m. on July 30. Shortly thereafter, Colonel Scott opened the hatch and stood on the top platform for about half an hour viewing the moonscape. He then reentered the cabin and closed the hatch after which the cabin was repressurized so that the astronauts could remove their spacesuits and rest before beginning their extravehicular activities.

Colonel Scott took his first step on the lunar surface at 9:26 a.m. on July 31, and in eight minutes was followed by Colonel Irwin.[3] They proceeded first to gather contingency samples of lunar soil and rocks and then to unload the lunar rover, a machine that would allow them to cover ten times more terrain than was possible on foot. They discovered that the front steering mechanism was inoperative and that they would have to rely on the steering mechanism of the rear wheels for maneuvering it over the rough lunar surface. They took off in the Rover at 11:20 a.m., after securely buckling themselves in it for the rough ride under conditions of the low gravitational force of the moon. *This marked the beginning of the motorized travel on the surface of the moon.*

The first stop, after traveling for about a mile, was at the rim of Elbow Crater, near the edge of Hadley Rille. There they activated the television camera mounted on the Rover to televise themselves gathering samples of the lunar soil and rocks and performing other activities. Some of the pictures revealed the inside of Saint George Crater, with slopes as much as thirty degrees, which was the farthest point reached on this excursion. Near St. George Crater, they examined a large boulder to which was attached a crystalline formation. From the heights near the crater the camera was focused on the walls and floor of Hadley Rille, some twelve hundred feet deep.

The Lunar Rover Parked Near the Lunar Excursion Module of Apollo 15 Looking Northward with Mount Hadley in the Background. Astronaut Irvin at the Rover. (Source: NASA - Photo No. 71-H-1413)

The two astronauts spent about two hours riding in the Rover and covered approximately five miles. Upon their return to the lunar-excursion module, Colonel Irwin deployed the ALSEP package of scientific instruments and Colonel Scott proceeded to drill two holes in the surface. In the first hole, about sixty-four inches deep, he placed a temperature sensor, but encountered difficulty in digging the second hole because of the bedrock. He had to postpone its completion until the next extravehicular activities. The excursion was shortened after Mission Control advised them that Colonel Scott was consuming oxygen at too high a rate. As a result of this, the total extravehicular activities, planned for seven hours, lasted only 6 hours and 34 minutes.

Before beginning the second extravehicular activities on August 1, at 7:47 a.m. a leak was discovered in the lunar-excursion module, caused by an unclosed valve in the urine dumping system. It was closed before enough oxygen had escaped to restrict duration of their stay on the moon. The astronauts seated and secured themselves in the Rover and traveled almost eight miles south across a field of craters, and rode several hundred feet up the gentle slopes of the Apennine Mountains, where they found crystalline rock. This rock was thought to be part of the original crust of the moon and probably some 4.6 billion years old. Following this motorized jaunt, they returned to the site of the lunar-excursion module and spent about three hours taking core samples of the lunar soil and rock, digging a trench, drilling in the soil, and erecting the American flag. The total time for their extravehicular activities amounted to 7 hours and 13 minutes, of which approximately four hours were consumed in driving about in the lunar Rover.

The last lunar excursion started early Monday morning, August 2, during which the astronauts covered 6.4 miles bringing the total mileage traversed in the Rover for all the excursions to 17½ miles. They drove to the Hadley Rille, taking 4 hours and 50 minutes, an hour less than planned. Before returning to the site of the lunar-excursion module, some difficulty was encountered in removing the ten-foot core tube which had been driven into the lunar surface to provide them with the deepest sample yet obtained from the subsoil structure of the moon. Colonel Scott also performed an interesting experiment to demonstrate Galileo's theory of falling objects of different weights dropped in a vacuum. He dropped a hammer and a feather from waist-level and the two objects reached the surface of the moon at the same time under the near-vacuum conditions and the low gravity force of the lunar environment despite the differences in their masses.

While Colonels Scott and Irwin were exploring the surface of the moon, Major Worden was conducting scientific experiments and taking photographs of the moon from the lunar-parking orbit. These experiments were dependent upon data being gathered by a variety of highly refined sensors, and were made possible because of the length of time spent in lunar orbit and the wide range covered by the cameras because of the altitude of the orbit. In addition, the length of the time spent in lunar orbit provided opportunities for making observations and photographing the moon under varying angles of sunlight. The results of these experiments should enable scientists to determine the location of minerals over extensive areas of the surface of the moon.

In the bay of the service module were 975 pounds of scientific instruments consisting of a scoop for capturing and analyzing traces of atmosphere if present in the near lunar environment within seventy miles of the moon's surface; a boom with a device for recording gamma rays with energies, characteristic of radio active substances; a device for identifying X-rays emitted by magnesium, aluminum, and silicon when exposed to the X-ray glares of the sun; and a device for registering volcanic activities when the spacecraft passed over deep fissures.

Colonels Scott and Irwin repressurized the cabin of the ascent stage of the lunar-excursion module at 9:42 a.m., on August 2, and shortly thereafter lifted off the surface of the moon. *The camera, left behind on the Rover, televised the lift-off, which was the first time this had ever been done.* The pictures did not show a fiery ascent due to the lack of oxygen but did indicate how rapidly the ascent stage rose because of the low gravity and lack of atmospheric friction. The rendezvous and docking of the ascent stage with the command-service module was accomplished without mishap. After docking, an oxygen leak was discovered in the hatches delaying the jettisoning the ascent stage of the lunar-excursion module for two hours. Resealing the hatches eliminated the trouble and permitted them to be secured.

On this mission, the Apollo 15 command-service module, after the separation of the ascent stage of the lunar-excursion module, continued to orbit the moon for two days in order to complete the experiments and photographic assignments conducted by Major Worden. On August 5, shortly before leaving the lunar orbit, a 78.5-pound satellite was released from the bay of the service module to orbit the moon for at least a year or longer, transmitting data relating to the moon's gravity and solar flares. Pictures were also taken of the dim light called, "Gegenschein" (reflected light) believed to be caused by particles of dust reflecting sunlight. It was anticipated that by comparing these pictures with those taken by astronauts on other Apollo missions, and others taken from the earth, there would be a better understanding of this phenomenon.

On August 4, the Apollo 15 spacecraft was inserted into an earth trajectory, ending six days of lunar exploration. Major Worden, clothed in his spacesuit, left the cabin of the command-service module at 11:40 a.m. on August 5, when the spacecraft was 196,000 miles from the earth, to retrieve film cassettes from the service module and to inspect certain scientific instruments. *He was the first man to carry out a floating extravehicular excursion in interplanetary space,* and was the ninth astronaut to leave a spaceship while in space. Assisted by handrails, he completed three trips to the rear of the spacecraft requiring 16 minutes and covering a distance of 15 to 20 feet.

The astronauts also took part in a 30-minute televised space-to-earth news conference on August 6, answering questions previously submitted to Mission Control and transmitted to them. In addition, they televised the eclipse of the moon, showing it in almost complete darkness, and carried on observations, in prearranged collaboration with the Soviet and Dutch astronomers of "black holes" in the sky.[4]

The Apollo 15 command module splashed down in the calm water of the Pacific Ocean at 4:46 p.m. (EDT) on August 7, carrying the astronauts who had exceeded the space endurance record set for all Apollo missions. Despite the malfunction of one of the parachutes, the capsule landed about three hundred and thirty-three miles north of Hawaii, one-half mile off its planned target and within view of the recovery ship, USS *Okinawa.* The only effect of the malfunctioning parachute was to increase the speed of descent from 28 to 32 feet per second. Within an hour after the splashdown, the astronauts were onboard the recovery ship where they received a telephone call from President Nixon. They did not have to undergo a three week quarantine period, but instead underwent four days of medical examinations. On August 8, they left the recovery ship to return to their homes.

The *Apollo 16* spacecraft blasted off at 12:45 p.m. April 16, 1972, on a scheduled 12-day return journey to the moon, and was the nation's fifth and next to last, lunar landing mission. It carried into space Captain John W. Young,[5] commander, Lieutenant Commander Thomas K. Mattingly, II,[6] pilot of the command-service module, the *Casper,* and Lieutenant Colonel Charles M. Duke, Jr., pilot of the lunar-excursion module, the *Orion.* The landing site was located on a plateau among the rugged mountains near Descrates Crater about one and a half miles above the site where Apollo 11 had landed. Three miles north of the site were the Smoky Mountains, rising 1,500 feet above the plateau Cayley Plains. Southward, about the same distance, was Stone Mountain, rising to the same altitude. It was assumed that the plateau was of volcanic origin and contained minerals different from those of the surrounding elevations, because of the intense heat to which the lava may have been exposed as it flowed from the moon's interior. Nearby were two medium-size craters North Ray and South Ray with what appeared to be bright rays radiating from their centers.

Considerable attention was given to the preservation of the health of astronauts and the physiological hazards encountered by them while living and working in space. In this instance, precautionary steps were taken, as a result of unexpected medical problems, which arose on the Apollo 15 mission, revealing that the

The Earth as Viewed from Apollo 16, about One Hour after Translunar Injection. Through the cloud cover are revealed a large part of the United States, most of Mexico, and part of Central America. (Source: NASA - Photo No. 72-H-635)

astronauts had experienced loss of potassium and heart-beat abnormalities. In that instance, the two astronauts who had landed on the moon lost 15 per cent whereas the one who had stayed in the lunar orbit lost only ten per cent.

The spacecraft assumed a 110 mile earth orbit within about twelve minutes after lift-off, and circled the earth one and a half times before being inserted into a lunar trajectory. Following this maneuver, the command-service module and the lunar-excursion module, with the third stage still attached, were linked nose-to-nose. Later, the third-stage rocket was jettisoned and placed in a separate lunar trajectory causing it to crash on the surface of the moon.[7]

Some concern arose when the astronauts observed places on the exterior of the lunar-excursion module where the insulation had peeled, believed to have been caused by the heat from a control thruster on the command module that had been fired excessively. After consultation, it was decided that this would not endanger the moon-landing mission and a midcourse correction was made at 7:33 p.m., when the spacecraft was approximately 136,500 miles from the earth and traveling at 3,000 miles an hour.

The astronauts spent their second day in space, April 18, mostly checking the systems for minor malfunctions. Commander Mattingly carried out an experiment to determine if molecules in liquids separate with greater purity under conditions of weightlessness. The experiment involved passing an electrical current through three transparent tubes filled with plastic particles suspended in a fluid. This caused the various sized particles to flow to the end of the tube opposite to that where the electrical current was introduced. The results of this experiment should have significant bearing upon the use of space stations as potential production centers for vaccines and medical products requiring extremely high levels of purity. Another experiment was conducted to observe the nature of the light flashes, with the astronauts blindfolded, as was carried out on other Apollo missions.

Shortly before entering the lunar orbit, a false computer signal temporarily locked the spacecraft's navigation alignment system, but it was decided that this would not materially interfere with the lunar-orbiting maneuver. On April 19, the spacecraft assumed a lunar orbit with an apolune of 195 miles on the near side of the moon and a perilune of 67 miles on the far side. After the spacecraft had circled the moon for four hours, the orbit was lowered to one with an apolune of 69 miles and a perilune of 12 miles in preparation for the separation of the lunar-excursion module and its descent at 1:08 p.m. to the surface of the moon.

On April 20 at 3:17 p.m. Mission Control delayed the landing of the lunar-excursion module, instructing the astronauts to remain in a station-keeping orbit until given the go or no-go signal. This was done to permit investigation of the extent of a malfunction in the secondary guidance system, causing some oscillation of the command-service module and preventing the circularization of its lunar orbit for carrying out its scientific and photographic activities. The delay lasted for six hours after which Mission Control gave the two astronauts the go-for-a-landing signal.

Threatening events continued to arouse concern as to the advisability of landing on the moon up to the time of the final descent. The steerable antenna on the lunar-excursion module, the main link between the two astronauts and the ground control station, locked in a fixed position. Although this problem was not resolved, the lunar landing was not cancelled because reliance could be placed on two other antennas. However, this deficiency did prevent televising the first lunar steps of the astronauts.

Trouble was encountered with the landing radar used for indicating the distance of the lunar-excursion module from the surface of the moon during its descent. There were also "funny" signals caused apparently by the radar beaming signals off the moon's surface. In spite of these malfunctions, the lunar-excursion module landed smoothly between the North Ray and South Ray craters within five hundred feet of the planned target site on April 20 at 9:23 p.m. Because of the delay in landing, the astronauts took an eight-hour rest period before starting their first extravehicular activities.

The first extravehicular activities began at 11:47 a.m. and ended at 6:58 p.m. on April 21 lasting 7 hours and 11 minutes. Captain Young stepped on the surface of the moon for the first time at 11:58 a.m. and, within two minutes, was followed by Colonel Duke.[8] Their feet sank into the greyish soil less than an inch, suggesting that it was not as thick as at the previous Apollo mission sites. Lunar rocks, some relatively large covered about twenty to thirty per cent of the surrounding terrain. About seventy per cent of the moon's surface in the region was pocked with small craters.

The astronauts unloaded Rover II and at first concluded that one of the batteries was dead but this conclusion proved unfounded. They then proceeded to set up the ultraviolet camera, deployed the high-gain antenna, activated the television camera on the Rover, collected samples of lunar soil and rocks, and erected the American flag near-by the landing site. By 1:30 p.m. they had traveled from two hundred to three hundred feet southwest to set up the scientific nuclear station with its series of experiments. Inadvertently, Captain Young got his foot tangled in the cable disconnecting equipment for a high priority experiment having to do with the measurement of the flow of heat from the interior of the moon. This necessitated discontinuance of the experiment.

During this extravehicular excursion, they put in place the first astronomical observatory for observing the celestial bodies from beyond the atmosphere of the earth. This miniature semi-automatic observatory had the capability of making photographic and spectrographic observations at ultraviolet light wavelengths which are unable to penetrate the earth's atmosphere. This was given one of the highest priorities because it was hoped that it would provide answers to the phenomenon of the "missing mass" in the space.[9] The highest priority was assigned to collecting samples of the lunar soil and the rocks in anticipation that they would reveal the composition of the formations located in this particular region.

The second excursion of six miles started at 11:33 a.m. and lasted until 6:56 p.m. on April 22, or 7 hours and 23 minutes in duration. Fifty minutes were spent near the landing site moving the ultraviolet camera out of the bright sunlight, taking panoramic pictures of the Descartes Crater, and loading Rover II with the tools and equipment. The first stop was Survey Ridge, a broad elevation at the top of a crater-pocked area strewn with boulders, apparently having been thrown out from the South Ray Crater. They then drove up the ten degree slope of Stone Mountain, also covered with boulders, and discovered that its magnetic field was of opposite polarity to that of the area near the landing site about three miles away.[10] Looking back they could see the lunar-excursion module from an elevation of 750 feet above the landing site. They discovered many sharp angular rocks (breccia) which surprised them because geologists had assumed that the region was so old that it would be covered with rounded boulders. It had also been assumed that crystalline rocks would be found relatively unchanged from their original state. They proceeded, in spite of the fact that the Rover's navigational system gave them trouble and the right rear fender fell off showering them with moon dust. Finally, they returned to the lunar-excursion module following their outward tracks.

The third and final extravehicular excursion initiated at 10:25 a.m. and ending at 4:05 p.m. on April 23, or 5 hours and 40 minutes in duration, had been reduced two hours because of the delay in landing the lunar-excursion module. The astronauts drove three miles to the rim of the North Ray Crater, about 3,000 feet in diameter arriving there at 11:40 a.m. and photographed the streaks of light rock-like debris radiating from its center. Many sharp-edged rocks, some about the size of a house, were found and samples of moon dust were taken by pressing velvet and spacesuit cloth against the lunar surface. They walked along the rim of the crater and down its inner side, sloping ten to 15 degrees where they discovered white rocks. Although half of the crater's interior was littered with boulders, some ten feet in diameter, they could not find bed rock. At another stop, they found crystalline rocks, milky in color, with many tiny holes, presumably formed by venting gases when the rocks cooled and hardened billions of years ago. The area explored was between two craters, North Ray and South Ray, about half a mile in diameter.

While these activities were taking place on the surface of the moon, Lieutenant Commander Mattingly was conducting a series of experiments from the command-service module as it circled the moon in a parking orbit. These experiments consisted of taking photographs of the moon and monitoring a series of instruments for gathering data relating to it and lunar space.

Captain Young and Colonel Duke returned to the lunar-excursion module and loaded 209 pounds of lunar soil and rock samples and other equipment into the ascent-stage cabin. They had spent approximately seventy-one hours on the moon, a record exceeding the 66½ hours spent by the astronauts of Apollo 15. The lift-off at 8:25 p.m. on April 23 was televised by the remotely controlled camera mounted on the Rover, left behind at the lunar landing site. Although this was the second televised lunar lift-off, *it was the first time that an ascent stage of the lunar-excursion module in lifting off the moon had been tracked by a television camera for at least two minutes. It was also the first time that a discarded descent stage of a lunar-excursion had been shown on television.* The television camera continued to operate for a brief period after the departure of the astronauts.

Shortly after its lift-off from the moon, at 10:35 p.m., the ascent stage of the lunar-excursion module was linked with the command-service module, piloted by Lieutenant Commander Mattingly, and continued to orbit the moon together. During the final hours in lunar orbit, an 85-pound subsatellite was released from the bay of the service module. It was expected that it would remain in lunar orbit for at least a year, transmitting data relating to the moon and the near-lunar environment. The ascent stage of the lunar-excursion module was jettisoned at 3:54 p.m. on April 24 and tumbled in space out of control, disrupting plans for it to crash on the moon to produce shock waves for recording by the seismographs on the lunar surface.[11]

During earthward journey, Lieutenant Commander Mattingly conducted extravehicular activities when the spacecraft was about 200,000 miles from earth. He first retrieved a 72-pound cassette from the panoramic camera and then a 20-pound cassette from the mapping camera located at the rear of the command-service module. During these activities, he was connected with the command cabin by an umbilical line providing him with oxygen and serving as a line of communications with

Colonel Duke, who stood in the open hatchway to receive the cassettes and stow them in the command cabin. The televised space walk took about an hour, including time required, by Lieutenant Commander Mattingly to carry out an experiment for determining the effects of unfiltered radiation on microbes. It consisted of a container with three small trays, each with 280 chambers containing temperature-sensitive ultraviolet measuring solutions and microorganisms. This microbial ecology evaluation device (MEED) was designed to measure the effects of space vacuum, weightlessness, and solar ultraviolet radiation on five strains of bacteria, fungi, and viruses when the open case was pointed toward the sun for ten minutes.

During their last day in space, the astronauts participated in a 20-minute space-to-ground news conference, while the spacecraft was 123,000 miles from earth. They presented their personal reactions relating to the great spatial adventure in which they were participants and talked about their surprises and discoveries. There was concern when an alarm indicated a failure of the command-service modules guidance and navigation systems but the alarm disappeared as suddenly as it appeared and no problem arose.

On April 27 at 2:45 p.m. the Apollo 16 capsule with the astronauts inside splashed down in the Pacific Ocean about two hundred and fifteen miles southeast of Christmas Island and three miles from the recovery ship, USS *Ticonderoga,* and within forty minutes the astronauts were onboard the recovery ship. They did not undergo a long period of quarantine and on the morning of April 29 were flown to Honolulu on their way to Houston, being scheduled to arrive there that night. They had brought back 214 pounds of lunar soil and rock samples and 12,466 feet of exposed film, including photographs of the moon taken on the surface and from lunar orbit, together with other scientific data.

The final moon landing spacecraft of the Apollo Program, Apollo 17, rose from its launching pad at Kennedy Space Center at 12:33 a.m. (EST), on December 7, 1972, after a delay of 2 hours and 40 minutes, caused by a pressurization problem in the third stage of the Saturn V rocket. The halt in the countdown came 30 seconds before the scheduled time of lift-off. *This was the first night-time launching of astronauts of the United States into space.* Captain Eugene A. Cernan, a veteran of previous space missions, was in command,[12] and was accompanied by Commander Ronald E. Evans, pilot of the command-service module, the *America,* and Dr. Harrison H. Schmitt, a civilian geologist and astronaut, pilot of the lunar-excursion module, the *Challenger.* This mission was planned for 12 days and 16 hours and was the most scientifically oriented of the Apollo Program. It had as its primary goal the determination of conditions under which men could live during extended periods of weightlessness in preparation for the coming space station (Skylab) program.[13]

The major threat during countdown was a labor dispute at the space center.[14] Fortunately, an agreement was reached between the parties one hour before the midnight strike deadline. None of the technical problems during the countdown proved unsurmountable. A switch was replaced in the Saturn V rocket, an emergency generator was replaced on the launching pad, a misfit filter was re-cut to size, and a leak was repaired in the water cooling system used during the electronic check.

The selected lunar-landing site for the Apollo 17 was a narrow valley on the southwestern edge of the Sea of Serenity called Taurus Littrow after the nearby Taurus Mountains and the Littrow Crater. It was the most eastern of all the Apollo landing sites and was chosen because it was rimmed by mountains rising between five thousand and seven thousand feet above the floor of the valley, mantled with dark material believed to be volcanic debris. It was hoped that at this site there would be discovered some of the oldest and the youngest lunar rocks.

Five brown and white mice (pocket mice), about the size of one's thumb, were carried in the spacecraft to serve as targets for cosmic ray particles. Beneath their scalps, radiation-sensitive films had been implanted to record the trajectories of cosmic ray impacts. The mice were to be flown to Samoa within six hours after splashdown and their brains examined for traces of radiation.

After two orbits of the earth, the third stage rocket was refired at 3:45 p.m. for six minutes plus a few seconds to increase the velocity of the spacecraft and to overcome the loss of time resulting from the launching delay. Following the linking together of the command-service module and the lunar-excursion module, the third stage was jettisoned and placed in trajectory causing it to crash on the surface of the moon.[15] The only problem encountered was a partially engaged docking latch, but this did not prove to be serious enough to delay the lunar landing.

Following the inspection of the lunar-excursion module, an experiment was carried out to determine how fluids behave under conditions of zero gravity. Commander Evans took a small transparent cylinder filled with a liquid in which suspended crystals changed color when heated and activated by an electrical heater at one end of the cylinder. Pictures were taken of the changing colors of the crystals, to indicate their reactions.

On December 10, at 2:47 p.m., the Apollo 17 spacecraft assumed a lunar orbit with an apolune of 194 miles and a perilune of 60.5 miles and circled the moon twice, after which the orbit was changed to one with an apolune of 68 miles and a perilune of 16.5 miles. The lunar-excursion module landed at 2:55 p.m., December 11, in a tilted position, about three hundred feet from the selected landing site near the edge of the Sea of Serenity partially surrounded by the North Massif and South Massif.

Captain Cernan stepped onto the lunar surface at 6:05 p.m., on December 11, and was followed within a short time by Dr. Schmitt. They assembled, loaded the Rover, and mounted on it a television camera. About four hours after landing they drove to the foot of a steep-walled mountain to the southwest of the landing site, taking time out to repair the rear fender of the Rover. They stopped to set up the nuclear scientific station about 300 feet from the lunar-excursion module. The station consisted of scientific devices to measure the flow of heat from the moon's interior, determine the moon's gravity variations, detect traces of gas, listen to the internal shock vibrations caused by setting off explosive charges, and observe micrometeorite impacts. At the foot of the steep-walled mountain, they found blue-grey and tan rocks and greyish soil containing fine grains of orange, red, and white particles. The extravehicular activities lasted 7 hours and 12 minutes and 11 seconds, including time required for depressurization of the cabin and for its repressurization.

The second lunar excursion took place on December 12 when the astronauts drove toward a steep bluff and then on to the 7,000 foot South Massif. Captain Cernan and Commander Evans traversed about four miles in the Rover, traveling about six miles an hour. They found light-grained rocks, some six to twelve feet in diameter, near the Camelot Crater and also crystalline rocks with fine grains of light color-reflective elongated crystals. In general, the rocks fell into two coloration groups: light-colored and very light tan or white. The blue-grey rocks found near the base of the mountain were coarse-grained conglomerate specimens appearing to be breccia. They also found orange and red soil along the rim of Shorty Crater. On this excursion, they covered about twelve miles and spent 7 hours, 37 minutes, and 22 seconds outside the lunar-excursion module, establishing a record.

The final extravehicular excursion started with the opening of the lunar-excursion module's hatch at 5:31 p.m. on December 13. The astronauts drove the Rover north to the base of the 8,000-foot North Massif where they discovered two boulders with blue-green crystals having glass-like veins which had apparently rolled down the mountain slope. They chipped pieces from the boulders, especially from the one with the blue-grey fragments, and took measurements of the gravity of the moon at that location.

On their return jaunt, they stopped at Van Serg Crater believing that it might prove to be a gas vent giving some indication of ancient volcanic action, but no such evidence was found. After returning to the site of the lunar-excursion module at about 12:30 a.m., they loaded about 250 pounds of samples of lunar rocks and soil together with equipment to be taken back to earth. They then held a ten-minute ceremony unveiling a plaque affixed to one of the legs of the descent stage of the

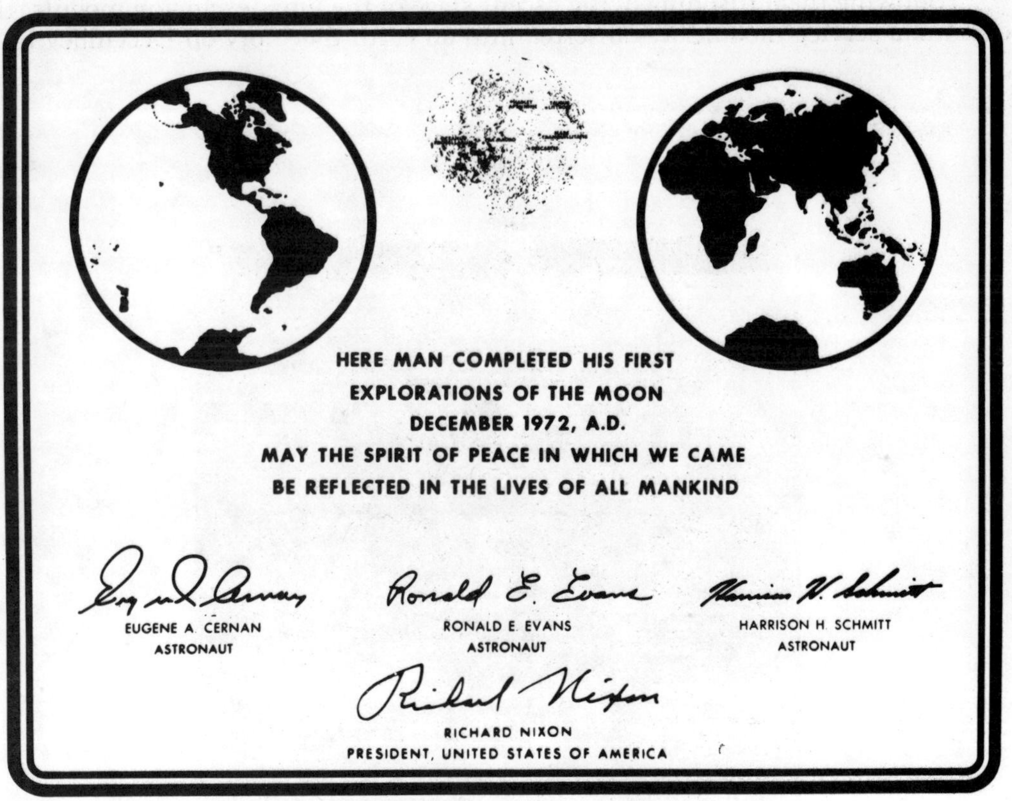

The Plaque Marking the Final Lunar Landing of the Apollo Program Left on the Moon. This plaque was unveiled at the end of the third extravehicular activity. It is made of stainless steel (9' x 7" x ⅝") and is attached to the ladder on the landing gear of the descent stage of the lunar excursion module. (Source: NASA - Photo No. 72-H-1541)

lunar-excursion module bearing the inscription: *Here man completed his first exploration of the Moon, December, 1972, A.D. May the spirit of peace in which we came be reflected in the lives of all mankind.* This plaque bore the signatures of President Nixon and the three astronauts of Apollo 17.

While Captain Cernan and Dr. Schmitt were engaged in exploring the surface of the moon, Commander Evans was probing the moon from the lunar-parking orbit with sensing instruments and cameras. The radio-sounding device was believed to be able to penetrate three-quarters of a mile into the moon's interior.

The ascent stage of the lunar-excursion module lifted off the moon at 5:56 p.m. on December 14, after a stay of 75 hours and 44 minutes on its surface. The television camera, mounted on the Rover, televised the lift-off of the ascent stage and followed its path upward for about 30 seconds or until it reached an altitude of about 1,500 feet. At 8:04 p.m., the ascent stage was docked with the command-service module, and with it orbited the moon for two days. During this time, the astronauts observed, from a distance of 75 miles what appeared to be orange soil at the rim of the Crater Sulpicus Gallus on the southeast edge of the Sea of Serenity directly across the plain from the landing site.

Following the jettisoning of the ascent stage of the lunar-excursion module, the command-service module was inserted into an earth trajectory on December 16 at

The Recovery Operation of Apollo 17 Showing the Apollo Capsule, the Recovery Ship, USS Ticonderoga, and the Recovery Helicopter (Source: NASA - Photo No. 72-H-1560)

6:35 p.m., after 75 orbits of the moon. The three astronauts were earthward bound. Colored pictures of the moon were transmitted during the early part of the return trip, when the spacecraft was 185,000 miles from earth. Commander Evans was televised performing a floating space walk to retrieve film cassettes from the experimental unit of the service module attached to the command module. The hatch of the cabin of the command module was opened at 3:33 p.m. and remained open until 4:20 p.m. During that time, he made three trips to the service module bringing back each time a film cassette which he gave to Dr. Schmitt standing in the hatchway. He also inspected the spacecraft's exterior and found places, where the silver insulation paint had blistered, and observed particles of ice around the nozzle of the valve of the water-waste dump.

The astronauts participated in a space-to-ground conference during the return trip and expressed their personal views with respect to the Apollo project and its future. This conference was conducted when the spacecraft was some 110,000 miles from earth and traveling at 3,600 miles an hour.

The Apollo 17 capsule, containing the three astronauts, splashed down in the Pacific Ocean at 2:25 p.m. on December 19, after a 12½ day journey to the surface of the moon and back. In less than an hour, the astronauts were rescued and flown by helicopter to the USS *Ticonderoga* where they underwent relatively short medical examinations. This was the final mission of the Apollo Program initiated in May 1961 by the late President John F. Kennedy. As stated by President Nixon, in congratulating the Apollo 17 astronauts, it was *the end of a most significant chapter in the history of human endeavor.*[16]

The Apollo astronauts had in all spent 80 hours walking and riding on the surface of the moon and had traveled more than sixty miles at six lunar-landing sites—a most remarkable achievement for the Space Program of the United States.

REFERENCES

1. It was discovered, prior to the lift-off, that the flotation collar for the command module to be used during the recovery stage of the mission, at splashdown, had been slashed by some unknown person. This was not considered serious enough to delay the departure of the spacecraft, because it could be replaced before the termination of the mission.
2. Colonel David R. Scott participated in *Gemini VIII Flight,* March 16, 1966 (p. 85), and *Apollo 9 Flight,* March 3, 1969 (p. 96).
3. They became the 7th and 8th men, respectively, to set foot upon the surface of the moon.
4. These "black holes" are conceived to be remnants of stars which have so shrunk that no light can leave or pass close to them. This is because the density of these "black holes" is believed to be so great and their gravity so powerful that light and radio waves from them are shifted permanently to the red or long-wave end of the spectrum and can never escape from them.
5. Captain John W. Young participated in Gemini III Flight, March 23, 1965 (p. 83); *Gemini X Flight,* July 18, 1966 (p. 86); and *Apollo 10 Flight,* May 18, 1969 (p. 97).
6. Lieutenant Commander Thomas K. Mattingly, II, was grounded because of exposure to German measles and had to drop out of the *Apollo 13 Flight,* April 11, 1970 (p. 109).

7. The spent third stage of the Saturn V rocket crashed on the moon on April 19, shortly after *Apollo 16* reached the vicinity of the moon.
8. They became the 9th and 10th men, respectively, to step upon the surface of the moon.
9. It is theorized that great clusters of galaxies observed in the heavens do not exist. Those clusters are moving at such fast rates of speed, relative to one another, that they would have long ago disintegrated, unless they were held together by powerful gravitational force. If they contain sufficient mass to generate such gravity, it must be thirty times more abundant even than can be seen in the galaxies themselves. This is the interpretation of the "missing mass."
10. On earth such situations are found where the rock formations ajoin others which cooled at a different time and in which the polarity of the earth's magnetism is reversed.
11. It was thought that it would ultimately drop out of its lunar orbit and crash upon the moon within a year or so, and its continuation in the lunar orbit would not present a hazard to future Apollo launchings.
12. Captain Eugene A. Cernan participated in *Gemini IX-A Flight,* June 3, 1966 (p. 86) and *Apollo 10 Flight,* May 18, 1969 (p. 97).
13. In total, the United States had logged more than eighty-six hundred man-hours in space for its manned space missions up to this time and had found that there were costs in terms of physical and psychological reactions on the part of the astronauts. It was evident that the changes were tolerable only if they leveled off at new equilibria. At what levels of tolerance or under what conditions were not known. Increased time and additional information were needed in order to explore more thoroughly these uncertainties.
14. A labor dispute between the Boeing Company and 60 space workers at the space center.
15. The spent third stage of the Saturn V rocket of Apollo 17 was scheduled to crash on the moon on Sunday, December 10, 1972.
16. Quoted in *The New York Times,* Wednesday, December 20, 1972 (1:1).

Chapter XI

EARLY SPECULATIONS RELATING TO EARTH-ORBITING SATELLITES AND SPACE STATION EFFORTS OF THE SOVIET UNION

> *Sceptics would do well to remember that saying that "the difficult takes some while and the impossible only a little longer"—fireworks into space rockets, for example, and changed techniques alter man's view.*[1]
>
> *H. E. Ross*

The placing of permanent man-made satellites or workshops in orbit about the earth is not a new concept but one which has gained in stature during the last decade. There have been fictitious accounts of earth orbiting man-made satellites and considerable scientific speculations with respect to them. In some respects, these speculations have become a reality and a new era of space exploration and travel has emerged.

There is a story by Edward Everett Hale, published during the latter part of the 19th century, about a fictitious brick moon hurled accidentally into an earth orbit by centrifugal force generated by whirling flywheels.[2] The principals in the story were Mr. Q. George Orcutt, Ben Haliburton, and Frederick Ingham, who, during their college days, deplored the fact that while the North Star served for determining latitude there was no natural star for determining longitude. They considered the possibility of building a brick moon and launching it into an orbit about the earth.

Some thirty years later, recalling their earlier contemplations, they initiated a project which materialized accidentally. Orcutt had amassed a fortune by investing in railroads and agreed to back the project financially and to provide the engineering knowledge for the development of the project. Mr. Ingham started a church and as founder and minister received a grant of land. Mr. Orcutt founded a school, became its first teacher, and also received a grant of land. It was later discovered that the land contained timber, clay suitable for bricks, and water power, all of which could be used in building and launching the 200-foot artificial satellite.

A dam was constructed to provide a waterfall to drive two giant flywheels below it, so placed that they revolved in opposite directions, to accumulate and store the energy released by the falling water. It was calculated that, by the time the artificial moon was completed, there would be sufficient centrifugal power, stored in the revolving flywheels, to hurl the man-made moon into an earth orbit. One flywheel was slightly smaller than the other which would cause the artificial moon, when hurled into space, to follow a path concentric to the curvature of the earth. The building site was just above the whirling flywheels, so that when completed the artificial moon could be rolled down the grooved side onto them.

Fortunately or unfortunately things did not work out as planned because the promoters of the brick moon did not anticipate that it would become a manned-satellite. The spring thaw softened the ground causing the supporting structure to give way so that the partially constructed brick moon slid onto the whirling wheels. It was hurled into space with tremendous force carrying with it George Orcutt and others, who, with their families, had moved into it during the cold winter months.

A year passed without any information about the brick moon and its occupants. A new star, Io Pheobe, was discovered arousing the curiosities of Ingham and Haliburton who began a long and arduous search for it. Eventually, they fixed its position and with their telescope discovered that it was made of brick and inhabited by human beings, who turned out to be their former friends. The brick moon was covered with green vegetation except the unfinished part appearing as a dark spot and serving as an entrance to its interior where its inhabitants lived. They also observed that the people performed a routine of twenty-foot leaps and hops at certain times each day, which proved to be a code for sending messages.

Haliburton devised a method of replying by laying large black letters made of cambric upon the snow-covered Saw Mill Flats. They could be seen clearly by George Orcutt from the artificial moon with the aid of his telescope, because there was little atmosphere about the satellite to distort the light and diffuse the objects. The task of arranging and rearranging the letters became so burdensome that Haliburton designed two carts to be moved in and out of a large building, upon which were placed large letters so arranged as to convey messages. In this manner, he speeded up the communication process.

Such science fiction did little to bring about actual travel in the exploration of space. This was left to scientists and technologists who through scientific speculation and research provided the know-how and techniques essential for the spectacular achievements in the exploration and exploitation of space during the Twentieth Century.

The Era of Scientific Speculations

Hermann Oberth Proposed Space Stations[3]

Hermann Oberth visualized several types of space stations. He proposed a fixed orbit space station to be assembled in an earth orbit 6.6 earth-radii distant from the center of the earth.[4] Its revolutions about the earth would be at the same speed as that of the earth when following this orbit, placing it above the same point on the earth's surface at all times. He suggested that such fixed orbit stations could be of military significance and used for launching bombs on earth targets.

He also proposed a springboard space station[5] to be used as a starting point for spaceships on exploratory expeditions. It would be placed in earth orbit on the fringe of the earth's atmosphere so that vehicles launched from it would not be slowed down by the atmosphere. The lower the orbit the greater would be the saving in the size and power of rocket ships to ferry men and materials, but such orbits would be unsatisfactory for surveillance of the earth by rocket ships. Its orbit would be in a west to east direction and in the equatorial zone, to take advantage of the earth's rotational speed and the low gravitational force in that zone.

Oberth contemplated the construction of a huge space mirror, measuring 60 miles in diameter and encompassing an area of 27,000 square miles.[6] The materials for constructing this space mirror would be carried into space by ferry spaceships and assembled in a low earth orbit to permit better access to supplies, materials, and personnel. Later it would be moved to a permanent orbit farther removed from the earth. The power to move the huge mirror would be obtained from the pressure exerted by the sunlight on the mirror's extensive surface. The light pressure would be controlled by manipulating the individual facets of the composite mirror and would be used to provide light for towns, melt icebergs, and regulate the weather and temperatures of extensive areas on the earth's surface. He estimated that it would take some ten to fifteen years to construct the space mirror but he was confident that it could be done.

He suggested also that a dwarf-asteroid five and a half to six miles in diameter could be used as a platform for an observatory instead of an artificial satellite.[7] Oberth theorized that a captive asteroid could be shifted into a proper earth orbit by rocket motors and kept there by periodical rocket firings.

Smith and Ross Proposed Orbital Space Station[8]

R. A. Smith and H. E. Ross gave considerable thought to a space station to consist of three major parts: a large metal mirror with a diameter of 200 feet to collect the energy of the sun; a bun-shaped body 100 feet in diameter, equipped with radio arrays and television cameras at one end; and a chamber laboratory with access to the station proper at the other end. Because of its complexity and size, the proposed space station, would be assembled in space after various sections had been prefabricated on earth and transported to the selected earth orbit by ferry spaceships.

The polished metal mirror, an essential part of the station, would be its source of power. It was estimated that it would be able to deliver to the piping located at its focus about 3,900 kilowatts which could be converted into a maximum of 1,000 kilowatts for operating the station's equipment and controlling its temperature.

The personnel would be located in the main body of the station divided into two concentric galleries surrounded by a central chamber. In the galleries would be quarters for personnel, amenities, laboratories, and workrooms, and in the hub compartments would be the air conditioning plant, radio gear, and electric motors. The galleries and the hub would be subjected to centrifugal force offsetting weightlessness by spinning the station. The outermost walls would be the floors and the innermost walls the ceilings, but the occupants would sense this as a normal up and down orientation.

Dr. Wernher von Braun Proposed Space Stations[9]

Dr. Wernher von Braun proposed a space station shaped like a huge spoked wheel. It would be constructed of nylon and plastic fiber, with a diameter of 250 feet and a peripheral rim 30 feet thick and would consist of 20 sections ferried from earth in a collapsed form and assembled in space. Once the sections had been put together, the whole structure would be inflated with air and component gases brought from earth in large collapsible tanks. A large highly polished parabolic metal mirror with piping at its focus would be placed along one side of the wheel-shaped space station, to provide energy for its mercury-fluid-turbo-generating system. The station would be painted white for temperature control supplemented with thermostatically controlled shutters with radiating surfaces. Its outer walls would be self-sealing as a guard against excessive loss of its interior air if penetrated by meteorites but also protected by a thin metal shield as a meteor bumper.

Within the space station itself, the effect of weightlessness would be offset by centrifugal force equal to one-third of the earth's gravitational force. The atmosphere inside the space station would be an oxygen-helium mixture kept at about normal pressure and the water requirements would be supplied by a water-recovery plant. Waterborne algae would be used to maintain the fundamental equation for proper air conditioning.

The food requirements were estimated at about four pounds per man-day and would consist largely of precooked and concentrated foods. The sewage, garbage and other wastes were estimated at about two tons per week, and would be carried back to earth or disposed of by small rockets loaded with the refuse and fired in an opposite direction to the orbital motion of the space station. These rockets would reenter the earth's atmosphere and disintegrate because of the intense frictional heat.

The space station would be placed in a polar orbit or near-polar orbit at about twelve degrees to the pole and at an altitude of 1,075 miles above the earth requiring two hours to complete at 15,840 miles an hour.[10] It would make 12 revolutions every 24 hours along its polar orbit and every part of the earth would be subjected to its surveillance. This orbit would be close enough to the earth to permit the economical and rapid construction of the space station, provide close military and meteorological observations, and serve as a way-station for moon and planetary journeys.

The Soviet Union's Space Station Efforts

An unmanned research workshop called Salyut-1 (Salute) was hurled into space on April 19, 1971, and assumed an earth orbit with an apogee of 138 miles and a perigee of 124 miles requiring 88.5 minutes to complete and inclined at 51.6 degrees to the equator.[11] The objectives of the mission were to test the elements of its onboard systems and to conduct scientific research and experiments. The placing of a space station in an earth orbit raised speculations as to whether cosmonauts would be ferried in a Soyuz spaceship to man it. In general, the Soviet scientists remained secretive and it was not until it was boarded by the crew of Soyuz XI in June that any information was made available relating to the mission.

The space station, powered by solar-battery panels, was reported, when linked with the Soyuz spaceship, to weigh around 55,000 pounds and to measure sixty-five feet long. The Soyuz spaceship weighed about 15,000 pounds putting the weight of Salyut-1 at about 40,000 pounds. It consisted of several cylindrical sections, all pressurized, ranging in diameter from six and a half to thirteen feet. Entrance to the space station was through a docking tunnel connecting with a small compartment containing equipment and instruments. A hatchway provided entrance to the main compartment with seats for four cosmonauts, which widened from an intermediate section to a cylindrical section 13 feet in diameter. At the end of the space station was the service module containing the fuel tanks and propulsion equipment.

Soyuz X was launched on April 23, 1971 at 2:54 a.m., Moscow time (the second night launching of a manned-spacecraft) four days after Salyut-1, with Colonel Vladimer A. Shatalov[12] in command accompanied by Aleksei S. Yeliseyev,[13] flight engineer and Nikolai Rukavishnikov, test engineer. The Soyuz X assumed an initial earth orbit with an apogee of 153 and a perigee of 129 miles and later succeeded in making a rendezvous and docked with Salyut-1. Reportedly, the Soyuz spaceship overtook the space station in two stages: the first, brought it within five hundred and ninety miles of the unmanned workshop and was carried out automatically; the second achieved rendezvous and docking and was carried out manually at 4:47 a.m. on April 24. After docking, the cosmonauts flew the spacecraft linked together for five and a half hours to test their improved systems in an earth orbit at an altitude of 150 miles. They circled the Salyut for approximately an hour, after undocking Soyuz X from the space station, and took photographs of it from different angles. Soyuz X returned to earth at 2:40 a.m. April 25, after two days in space, landing about seventy-five miles northwest of the city of Karaganda in the Soviet Republic of Kazakhstan. According to Soviet space authorities, the mission accomplished its major purpose of testing the new rendezvous and docking procedures between a small manned spacecraft and a large unmanned earth-orbital space station.

The launching of Soyuz XI on June 6, 1971, at 7:55 a.m. Moscow time into an earth orbit was surrounded with the customary secrecy. Its crew consisted of Lieutenant Colonel Georgi T. Doblovolsky, Vladislov N. Volkov and Viktor I. Palsayev, the latter two being civilian engineers. *On June 7, Soyuz XI made a rendezvous and docked with Salyut-1 as it was making its 779th revolution of the earth and the cosmonauts went onboard, thereby establishing the first manned earth-orbital space station.* At the time, Salyut-1 was circling the earth in an orbit with an apogee of 155 miles and a perigee of 132 miles, inclined at 51.6 degrees to the earth's equator taking 88.2 minutes to complete. Its earth orbit was raised later to prolong its stay in space and reduce the drag caused by the rarified atmosphere.[14]

The three cosmonauts conducted scientific experiments and studied the effects of weightlessness while in the space station. They completed their first week aboard Salyut-1 on June 13, and on the following day, carried out an experiment involving a jet airliner and a biplane, to determine the physical state of the earth. During the experiment, the jet airliner flew at an altitude of five miles and the biplane at 1,000 feet to obtain a spectra reading of the earth's formation. At the time, the space station was in an earth orbit with an apogee of 172 miles and a perigee of 158 miles requiring 89.6 minutes to complete. On June 18, an experiment was initiated to test the effects of extended exposure to space conditions on the telescope mounted on the orbital observatory and its operation.

The cosmonauts held the first birthday party in space, on June 19, to celebrate the thirty-eighth birthday of Viktor I. Palsayev, and on June 22, showed television pictures of their space garden, an experiment having to do with the reaction of plant life to weightlessness. The first sprout appeared two days after the container was activated but the second, appearing later, grew faster.

The Soyuz XI cosmonauts exceeded the space endurance record on June 24, but there was no indication as to the duration of the mission. Salyut-1 had made more than 1,000 orbits of the earth and the cosmonauts had been aboard it for about 270 of them. June 28 was the cosmonauts twenty-third day in space, setting a new space endurance record five days longer than the previous record. On June 29, they transferred from Salyut-1 to Soyuz XI, tested its onboard systems, and unlinked the spacecraft without difficulty. Later, they reentered the earth's atmosphere with all systems operating normally. The braking maneuver was carried out, the parachute system released, and the spaceship landed smoothly within a pre-set area. When the hatch of the capsule was opened, it was discovered that *the three cosmonauts were dead.* Later investigation revealed that Soyuz XI's hatch had sprung a leak while plunging through the earth's atmosphere, allowing the oxygen to escape from the cabin of the capsule. This was determined to be the cause of the deaths of the cosmonauts who at the time were not wearing spacesuits.[15] They had established a space endurance record of 23 days, 18 hours and 22 minutes on this fateful mission.

On April 3, 1973, the Soviet Union launched Salyut-2, an unmanned earth-orbital space station. It was three times the size of Salyut-1, and designed to support three separate crews for a total of twenty weeks during an eight month period in space. Tass reported that Salyut-2 had been placed in an initial low earth orbit with an apogee of 161 miles and a perigee of 134 miles, the angle of inclination being 51.6 degrees to the equator. On April 11, it was confirmed by Soviet officials that its earth orbit had been raised on April 4 and 8, giving it an apogee of 184 miles and a perigee of 162 miles, an orbit not normally suitable for manned earth-orbital flights. The Soviet authorities did not indicate that it was their intention to man the space station, to rendezvous or to dock with it. General Vladimir Shatalov, Chief of the training project, explained that Salyuz-2 would yield valuable information with respect to weather forecasting, geology, transport, communication, forestry, agriculture and environmental protection, but did not intimate that a Soyuz spaceship would ferry cosmonauts for the purpose of manning the space station.

As early as April 25, some American space experts suspected that Salyut-2 had met with a catastrophe sometime during the previous weeks. This suspicion was based on the detection of some twenty-five fragments near its orbit, suggesting that there might have been an explosion. Furthermore, non-Soviet sources reported that the space station's radio signals had stopped and the space station was tumbling out of control in space. No statement was issued by Soviet authorities at the time, but on April 28, Tass reported that the Salyut-2 mission had been terminated and the data obtained from it would be used in building a new space station. However, no mention was made of its fate. The only evidence that it had broken up was based on observations and assumption by Western space observers.

Interest in manned earth-orbital flights was revived during the latter part of 1973. It was believed that Soviet space authorities were endeavoring to demonstrate renewed confidence in its manned space program. In addition, it was thought that they were seeking to offset criticisms by some Western space observers that they would be unable to carry out their part of the joint mission scheduled for July, 1975.

On September 27, 1973, the Soviet Union launched an earth-orbiting manned spaceship, the first in more than two years. Soyuz XII, bearing cosmonauts Lieutenant Colonel Vasily G. Lazarev and Oleg G. Makarov, was hurled into space on a two-day mission. This was *the first time that Soviet space officials had announced the duration of a flight in advance.* The purpose of the mission was to check out modifications made in the Soyuz spacecraft in converting it from a three-man to a two-man vehicle.

The spacecraft's initial orbit was changed by raising it sixty to eighty miles, and at the end of the fourth orbit it was reported to be following a path about the earth with an apogee of 155 miles and a perigee of 120 miles. Later, this was altered to an earth orbit with an apogee of 214 miles and a perigee of 202 miles. The flight of Soyuz XII was terminated on September 29, 1973, when the space vehicle with the two cosmonauts inside soft-landed in a predetermined zone near Karaganda in Kazakhstan about two hundred and fifty miles southwest of Baikonur Space Center. On October 2,

1973, high honors were bestowed on Lieutenant Colonel Vasily G. Lazarev and Cosmonaut Oleg G. Makarov, making them Heroes of the Soviet Union and awarding them the Order of Lenin and Gold Star Medals.

On December 18, 1973, Soyuz XIII was blasted into an earth orbit. In command was Major Pyotr Klimuk accompanied by Valentin Lebedev flight engineer; for both it was their first flight into space. No announcement of the space flight was made until about eighty minutes after lift-off at which time the two cosmonauts were shown in space on Moscow television. During this flight, the United States' Skylab 4 was in progress, *marking the first time that the Soviet Union and the United States had manned space flights in earth orbit simultaneously.*[16]

The Soviet space officials gave no indication as to the length of the mission, but during the fifth orbit of the mission, the spaceship was put into a near-circular earth orbit with an apogee of 169 miles and a perigee of 140 miles, inclined at 51.6 degrees to the equator taking 89.6 minutes to complete. Because in the past that type of earth orbit had been used for linking spacecraft, speculation arose on the part of some Western experts that the Soviets were planning rendezvous and/or docking maneuvers and that another spaceship or spaceships might be launched shortly. The cosmonauts continued experimentations which included studies of blood circulation in the brain under conditions of weightlessness, an onboard greenhouse for biological micro-organisms, testing the mechanisms for autonomous navigation under different flight conditions, X-ray photography of the sun, telescopic spectrography of stars in the ultraviolet range and spectrography of the features of the earth.

On December 20, the cosmonauts began their 29th earth orbit but still no indication was given as to the duration of the flight. It was announced on December 24 that the mission was nearing its end and on December 26 it was terminated. Soyuz XIII, with the two cosmonauts inside, soft-landed after eight days in space, and 128 orbits of the earth. The landing was successfully carried out during high winds and a snowstorm about one hundred and twenty-five miles southwest of Karaganda in the Kazakhstan Soviet Republic in Central Asia.

It was not until June 25, 1974 that the Soviet Union launched the space station, Salyut-3, similar to the one launched on April 19, 1971. It consisted of a double cylinder and a cone making three rooms, one of which was for recreational activities where the cosmonauts rested, ate, and carried on gymnastic exercises. The unmanned space station was placed in an earth orbit with an apogee of 167 miles and a perigee of 136 miles, inclined at an angle of 51.6 degrees to the equator, requiring 89.1 minutes to complete. The purpose of the mission was to continue the improvement of the design of the space station's sytems and equipment and conduct scientific experiments.

On July 3, 1974 at 9:51 p.m. Moscow time, Soyuz XIV was hurled into space with Colonel Pavel Popovich, a veteran cosmonaut,[17] in command, and Lieutenant Colonel Yuri Artyukhin, flight engineer, making his first flight into space. It entered an initial orbit with an apogee of 167 miles and a perigee of 136 miles inclined at 51.6 degrees to the equator, taking 89.1 minutes to complete. Later, it assumed a near-circular earth orbit with an apogee of 171 miles and a perigee of 138 miles inclined at 51.6 degrees to the equator, requiring 89.7 minutes to complete. For the first time tracking ships at sea were employed, making it possible for the Soviet Control Mission to be in contact with the cosmonauts at all times during the flight.

Before dawn of July 5, the two cosmonauts achieved a soft-linkage with Salyut-3 after pursuing it for 2,200 miles and catching up with it within twenty-six hours after lift-off. Soyuz XIV was guided automatically to within one hundred yards of the space station and then Colonel Popovich manually controlled it during the remaining distance to the space station.

At 4:30 a.m. Moscow time on July 5, Lieutenant Colonel Artyukhin entered Salyut-3 and was soon followed by Colonel Popovich. By 4 p.m. they had completed ten orbits of the earth aboard the space station in an orbit with an apogee of 171 miles and a perigee of 158 miles, inclined at 51.6 to the equator, taking 89.1 minutes to complete. There was no indication of the duration of the mission.

The experiments conducted onboard the space station included studies of the earth's surface, atmospheric formations, and the physical characteristics of space. In addition, they carried on medicobiological experiments and tested the improved design of the space station. An apparent purpose of the mission was to test the docking devices for the scheduled Soviet-Soyuz — United States-Apollo joint mission.

The Soviet space officials announced on July 15, that consideration was being given to the curtailment of the mission because of intense solar flares which had been severe during the first four days of the flight but had subsided. Fortunately, the flight path of Salyut-3 was within the earth's magnetosphere, otherwise the cosmonauts would have been subjected to solar radiation beyond the permissible norms.

On July 18, the cosmonauts moved the equipment and documents into Soyuz XIV and the next day tested its systems in preparation for the return trip to earth. They appeared on television to demonstrate that vibration in space could adversely affect the accuracy of instruments based on pendulum action. The 15-day mission was terminated on July 19, when Soyuz XIV, bearing Colonel Popovich and Lieutenant Artyukhin, soft-landed on target in the Soviet Republic of Kazakhstan in Central Asia. During their stay aboard the Salyut-3 they had photographed large sections of Soviet Central Asia, as a basis for identifying the locations of mineral deposits.

On August 26, 1974, at 10:58 p.m. Moscow time, the Soviet Union launched Soyuz XV carrying Lieutenant Colonel Gennady N. Sarafanov, commander and Lev S. Demin, a civilian flight engineer. Speculation arose among Western space specialists as to the possibility of the cosmonauts rendezvousing and docking with Salyut-3, the Soviet space station which had been circling the earth under automatic control since it was vacated by the Soyuz XIV cosmonauts on July 18, 1974. As usual little information was released by Soviet space authorities as to the purpose and duration of the mission.

By 5:00 p.m., Soyuz XV had completed 12 orbits of the earth and was reported to be following an earth orbit with an apogee of 172 miles and a perigee of 159 miles. It was inclined at 51.6 degrees to the earth's equator and required 89.6 minutes to complete.

Unexpectedly on August 28, it was announced that the Soviet cosmonauts were returning to earth without having docked with Salyut-3. It was claimed that the cosmonauts had made many approaches to the space station and had made observations and inspected it. It was further stated that the cosmonauts had conducted experiments to improve the techniques of piloting the spaceship under varying flight situations.

Western space specialists had assumed that the cosmonauts would board Salyut-3 and remain aboard for at least a week or ten days, since the Soyuz XIV crew had restocked the space station with food presumably for the crew of another mission. It was also assumed that they were testing techniques in preparation for the joint Soviet Union and the United States space venture set for July 15, 1975.[18] The unexpected termination of the flight left them perplexed.

The Soyuz XV soft-landed at night, under unfavorable conditions, 30 miles southwest of Tselinograd in the Soviet Republic of Kazakhstan at 11:10 p.m., Moscow time on August 28, after having been in space for only 48 hours and 12 minutes. It was thought by some Western space experts that because of the difficult night landing of the spaceship some sort of emergency had arisen, maybe the exhaustion of fuel in several attempts to dock with the space station. The cosmonauts were rescued shortly after landing and were reported to be in good health.

The Soviet Union sent Soyuz XVI into space on December 2, 1974 at 12:40 p.m. Moscow time with the declared mission of testing the onboard systems which had been modernized to meet the requirements of the joint Soviet Union and the United States, Soyuz-Apollo flight, scheduled for July, 1975. In addition, it had as part of its undertaking the assignment to make observations of sections of the earth's surface as a basis for information which might aid in the solution of the nation's economic difficulties. Lieutenant Colonel Anatloy V. Filipchenko[19] was in command of the flight and accompanying him was flight engineer Nikolai N. Rukavishnikov, a civilian.[20] These cosmonauts were the back-up crew for the joint Soyuz-Apollo flight.

It was reported that after assuming an earth orbit on December 3, the cosmonauts reduced the cabin atmospheric pressure from 14.7 pounds per square inch to

A View of the Soviet Soyuz Spacecraft in Earth Orbit Photographed from the American Apollo Spacecraft during the Joint United States/Soviet Union Mission, July 1975 (Source: NASA - Photo No. 75-H-891)

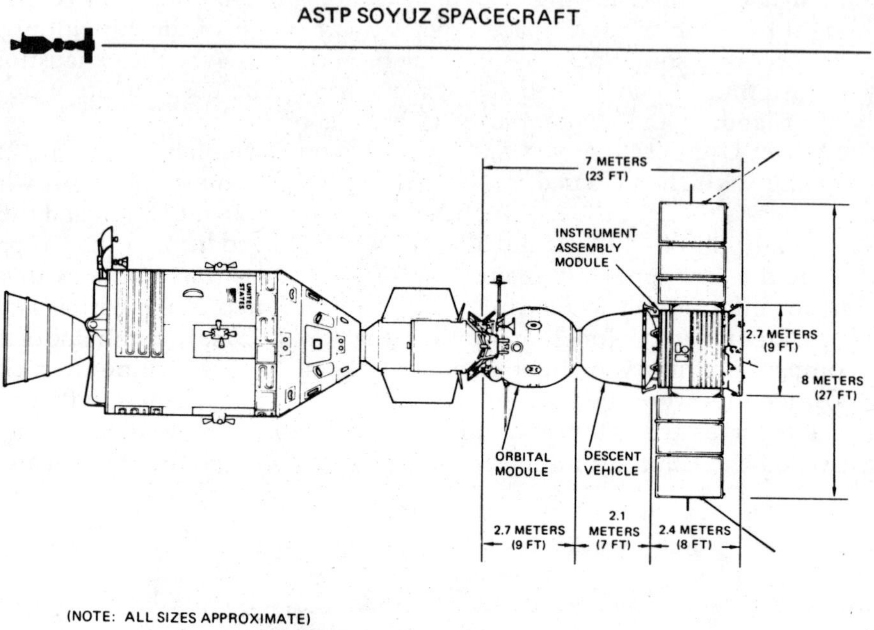

Drawing Showing Sections and Approximate Dimensions of the Soyuz Spacecraft which Participated in the Joint Mission of the United States and the Soviet Union during July 1975 (Source: NASA - Photo No. 74-H-726)

ten pounds per square inch and at the same time doubled the oxygen intake by raising it from 20 to 40 per cent. This was done to make the cabin atmospheric conditions closer to those which would prevail in the cabin of the Apollo spacecraft to be used in the joint flight by the American astronauts; that is an atmosphere of pure oxygen at a pressure of five pounds per square inch. The Soviet space officials use a mixture of nitrogen and oxygen and the adjustment was made to test conditions which would permit more rapid acclimatization by the cosmonauts in passing through the docking module into the Apollo spacecraft during the scheduled joint flight. They also carried out tests relating to the docking mechanism which were reported to have been performed successfully using a simulated ring, with the docking module's dimensions, mounted outside Soyuz XVI's own docking mechanism.

It was reported that during the sixty-eighth and sixty-ninth orbits, the cosmonauts photographed a panorama of the horizon, using polarized light over a flight path of nearly 20,000 miles. They also photographed a starlit part of the sky in checking out equipment to be used in the joint flight in July 1975, in which one of the spacecraft would pass between the other and the sun simulating a solar eclipse.

On December 7, Soyuz XVI had completed 81 orbits of the earth, most of which were completed at an altitude of 155 miles above the earth. This is the orbit in which the rendezvous between the Soyuz and Apollo spacecraft was to be programmed. No attempt was made by the cosmonauts to rendezvous or dock with Salyut-3 in orbit about the earth.

Before landing, the cosmonauts reestablished the cabin pessure at 14.7 pounds per square inch and adjusted its contents to the previous mixture of nitrogen and oxygen. On December 8, after six days in earth orbit, Soyuz XVI returned to earth, having completed 96 revolutions of the earth. It landed in the snow-covered region of Central Kazakhstan about one hundred and ninety miles north of Dzhezhazgan and about fourteen hundred miles southeast of Moscow. The landing was described as routine with both cosmonauts feeling well.

In connection with this flight, the Soviet space authorities were less secretive. They disclosed more information in advance about the mission, yet the television viewers were not permitted to view live the blast-off or the landing of the spaceship. It was assumed by some western space analysts that the flight was an endeavor on the part of the Soviet Union to demonstrate its professionalism in space and to counteract American concern as to its ability to carry out its part of the scheduled joint venture.

On December 26, 1974, the Soviet Union placed in earth orbit another space station, Salyut-4.[21] The high altitude earth orbit in which it was positioned gave it longer staying capabilities in orbit, permitting it to be manned by several different crews of cosmonauts at different times. It was described as an earth-orbital research laboratory shaped like a giant winged telescope with three major compartments and weighing about twenty tons. As of January 6, 1975, it was reported to have completed 180 orbits of the earth with its systems functioning normally.

On January 11, 1975 at 12:43 a.m. Moscow time, Soyuz XVII was sent into space carrying two space rookies, Air Force Lieutenant Colonel Aleksei Gubarev, commander and civilian engineer Georgi Grechko. The lift-off was from the Baikonur Space Center in Soviet Central Asia. The purpose of the mission was to conduct joint experiments with Salyut-4, presumably requiring rendezvous and docking as well as boarding the space station. The Soyuz spacecraft was reported to be circling the earth in an orbit varying between 221 and 183 miles, inclined at 51.6 degrees to the earth's equator.

Approximately thirty-three hours after blasting off, the cosmonauts brought the spacecraft to a successful docking with Salyut-4 and went aboard. The Soyuz XVII spacecraft was guided automatically during the first phase of rendezvous and docking, but the last 100 yards were carried out manually by the cosmonauts. The docking was completed while the spacecraft was out of contact with ground control stations. This was the first successful docking with a space station in earth orbit since the Soyuz XIV spacecraft linked with Salyut-3 during July, 1974. The cosmonauts after boarding underwent a strenuous schedule of putting things in order. This may have been the cause of their early difficulties in adapting to conditions of weightlessness.

During their stay in the space station the cosmonauts carried on a wide range of experiments including biological and medical experiments, earth observations and tests relating to improved design of the space station. In addition, they carried out complex navigational experiments resulting in new procedures for navigation with improved equipment permitting station orientation by the sun, moon, and planets under varying flight conditions. They also perfected a device for determining their position as they passed over the night side of the earth. The cosmonauts set up a teletype link with ground stations and most of their instructions were conveyed to

The Space Orbital Station, Salyut, Shown in the Cosmos Pavilion at the USSR Economic Achievements Exhibition. There are 1300 pieces of equipment in the Salyut. It is 14m long and weighs 18900 kg. (Source: TASS from Sovfoto — Photo No. 890257)

them by radio teletype. *This was the first time that Soviet space officials had established a space based teletype transmitting station.*

The two cosmonauts, on February 2, set a new Soviet space endurance record in an earth-orbital space station. They surpassed the 22½ days spent aboard Salyut-1 by the three cosmonauts of Soyuz XI who lost their lives while returning to earth in 1971.[22] No indication was given as to the duration of the Soyuz XVII mission.

On February 9, the Soyuz XVII mission was terminated and the two cosmonauts were returned to earth safely. The landing was accomplished under hazardous meteorological conditions. There was a heavy overcast reducing visibility to 500 yards and wind velocity was 40 miles an hour. In spite of these unfavorable conditions, they landed about 70 miles from Iselinograd in Soviet Central Asia successfully. They had spent 30 days in space, 28 of them aboard the earth-orbital research station, Salyut-4, setting a new Soviet space endurance record for manned orbital flights.

It was disclosed by Soviet space officials on April 7, 1975 that a manned Soyuz spacecraft launched on April 5, had been aborted shortly after lift-off and the two cosmonauts returned uninjured to earth. The two cosmonauts, Vasily Lazarev and Oleg Makarov soft-landed southwest of Gorno-Altaisk in Western Siberia. The intended purpose of the mission was to carry out joint experiments with Salyut-4 and was stated to have no connection with the joint flight scheduled in July.

The explanation for the failure of this undesignated Soyuz mission was the use of an old model carrier rocket. The third stage stretched the parameters of the carrier vehicle's movements, causing them to deviate from the pre-set values. An automatic device signaled the discontinuane of the flight, the detachment of the space capsule, and its return to earth.

The Soyuz XVIII spacecraft was launched on May 24, 1975 under the command of Lieutenant Colonel Pyotr I. Klimuk[23], accompanied by Vitaly I. Sevastyanov,[24] civilian flight engineer. Both cosmonauts had made previous space flights and had been designated Heroes of the Soviet Union. The lift-off took place at 5:58 p.m. Moscow time and the cosmonauts began docking procedures at 9:11 p.m. Moscow time on May 25. Shortly thereafter, they achieved a smooth docking in the dark.[25] *This was the first known time that a Soviet space station had been occupied in succession by cosmonauts while in earth orbit.* Several attempts had been made to do so. At the time of boarding by the Soyuz XVIII cosmonauts, Salyut-4, which had been orbiting the earth under automatic control after having been vacated by the Soyuz XVII cosmonauts on February 9, had completed 2,379 orbits.

The purposes of the mission were to test elements and systems of the spacecraft and seek uses for individual and/or groups of spacecraft for achieving scientific and applied undertakings in near-earth orbits. Experiments involving studies of the sun, planets, and stars were to be undertaken in the various ranges of the spectrums of electro-magnetic radiation, medical-biological research and test of the space station.

No indication was given as to the duration of the mission which came as a surprise to most Western space observers because of its nearness to the joint Soyuz-Apollo mission scheduled for July 15. The linked spacecraft were circling the earth in an orbit with an apogee of 221 miles and a perigee of 214 miles indicative of a long duration flight. As of June 13, the cosmonauts had spent 21 days in space and on the day of the launching of the Soyuz and Apollo spacecraft, participating in the joint international flight on July 15, they had reached their fifty-second day.

On July 26, Soyuz XVIII touched down at 5:18 a.m. Moscow time, 35 miles northeast of Arkalyk in Central Kazakhstan after having been in space 63 days. The cosmonauts were assisted from the space capsule apparently in good physical condition. The landing, in this instance, was not televised live as in the case of the Soyuz XIX spacecraft at the termination of the joint international flight.

The Soyuz XVIII mission was probably the most successful in the Soviet Orbital Laboratory Program. The cosmonauts, while in Salyut-4 carried out a heavy schedule of experiments including studies of the sun, stars, outer space, and the earth's atmosphere, and made an intensive survey of some three million square miles of Soviet territory. In addition, they raised peas and onions in a small greenhouse to determine the possibility of growing food in space.

ADDENDUM

On November 17, 1975, the Soviet Union put into earth orbit unmanned Soyuz XX which docked automatically with Salyut 4 on November 19. On June 22, 1976, it placed Salyut 5, a new space station, in earth orbit followed by Soyuz XXI on July 6, carrying Colonel Boris Volynov and Lieutenant Vitaly Zholobov. The spacecraft docked with Salyut 5 on July 7 and the cosmonauts went aboard the space station. No indication was given as to the duration of the mission. (See pages 180-184 for the Soyuz XIX mission.)

REFERENCES

1. Cited in Kenneth W. Gatland, *Prospect Satellite,* (London: Allan Wingate, 1958), Chapter 3, "Orbital Bases," by H. E. Ross, p. 108.
2. Edward Everette Hale, "The Brick Moon," *The Atlantic Monthly,* Vol. XXIV, 1869, pp. 451-469; 603-611; 679-687; Vol. XXV (1870), pp. 215-222.

3. Hermann Oberth, *Man Into Space, New Prospect for Man and Travel,* (New York: Harper and Brothers, 1957). (Translation)
4. *Ibid.*, p. 74.
5. *Ibid.*, pp. 73-74.
6. *Ibid.*, pp. 97-109.
7. *Ibid.*, pp. 70-71.
8. Kenneth W. Gatland, *op. cit.*
9. *Ibid.*, pp. 123-125; von Braun's proposal was originally published in *Collier Magazine* and later in England by Sigwick and Jackson, 1952, *Across the Space Frontier*; See: Lester Del Rey, *Rockets Through Space,* (Philadelphia: The John C. Winston Co., 1957) and Wernher von Braun, *Conquest of the Moon,* (New York: The Viking Press, 1953), Chapter 2, pp. 7-14 and Kenneth Gatland, *op. cit.*, pp. 123-126.
10. Wernher von Braun, *Conquest of the Moon,* (New York: The Viking Press, 1953), p. 13.
11. It was believed by some Western space experts that the Soviet Union launched a space station in December, 1970 but it was not publicly announced as a space station, but was designated as Cosmos 382, an overall designation for Soviet space shots. (*The New York Times,* Wednesday, June 26, 1974) (13:1).
12. Colonel Shatalov participated in *Soyuz IV Flight,* January 14, 1969 (p. 18), and *Soyuz VIII Flight,* October 13, 1969 (p. 19).
13. Mr. Yeliseyev participated in *Soyuz V Flight,* January 15, 1969 (p. 18), and *Soyuz VIII Flight,* October 13, 1969 (p. 19).
14. The unmanned Salyut-1 remained in earth orbit until October, 1971, when it was reported to have reentered the earth's atmosphere and disintegrated.
15. They were the 2nd, 3rd, and 4th cosmonauts reported to have died in the manned space missions of the Soviet Union. Vladmir M. Komarov, the first, was fatally injured on April 24, 1967, when Soyuz I spacecraft crashed on landing because its main parachute snarled (pp. 15-16).
16. At that time, the United States' Skylab-4 Mission was in an earth orbit (pp. 160-164).
17. *Vostok IV Flight,* August 12, 1962 (p. 12).
18. This joint international flight was successfully initiated on July 15, 1975 (pp. 180-184).
19. Lieutenant Colonel Anatoly V. Filipchenko was in command of *Soyuz VII Flight,* October 12, 1969 (p. 19).
20. Nikolai N. Rukavishnikov, test engineer, *Soyuz X Flight,* April 23, 1971 (p. 137).
21. It was reported by Tass on January 11, 1975 that Salyut-3, launched on June 25, 1974 had ended its mission.
22. *Soyuz XI Flight,* June 6, 1971 (p. 137).
23. *Soyuz XIII Flight,* December 18, 1973 (p. 139).
24. *Soyuz IX Flight,* June 1, 1970 (p. 22).
25. Both Salyut-4 space station and the Soyuz XVIII spacecraft were in the sun's shadow. The search lights of the Soyuz XVIII spacecraft were unable to penetrate the intense darkness.

Chapter XII
THE SKYLAB MISSIONS AND THE SPACE SHUTTLE PROGRAM OF THE UNITED STATES

Following the termination of the Apollo Program, efforts were made to develop, design and implement the space station (Skylab) and the integrated space transportation (Shuttle Spacecraft) programs. Planning for these programs began several years prior to the completion of the Apollo missions but because of drastic budgetary restraints increased reliance was placed upon the applications of Apollo hardware for the more immediate short-run manned space projects planned for the post-Apollo period.[1]

Space Station

Space stations have received much attention and have been the subject of several inquiries sponsored by the National Aeronautics and Space Administration. The general opinion of those directly involved in the future space activities of the United States is that space stations will be the key to the effective exploration and exploitation of space in the coming decades.

Three basic configurations relating to space stations were considered.[2] The first consisted of one or more independent three- to five-man cylindrical stations having a diameter of 22 feet. It would have the capability of docking with another station while in orbit or being stacked with other similar stations on earth and being launched as a unit into an earth orbit. The second was a highly integrated station capable of housing a crew of three to nine men. Its pressurized compartment would serve as a cabin, a work area, and for cargo transfers. The third consisted of a combination of multi-systems, crew, logistics, and experimental modules. This composite space station would provide flexibility through repeated launchings of manned and/or unmanned modules to adapt it to the diversified requirements of different nations.[3]

The Apollo Applications Workshop concept was basic in planning the initial American space station. The general idea was to change the upper stage of the Saturn V carrier rocket into a habitable space module. An airlock was to be inserted between the Apollo spacecraft and modified S-IVB rocket stage, which would provide the needed electrical power and the environmental systems. In addition, a tunnel was to be added to enable the astronauts to move between the two compartments as well as a porthole to permit their exit for extravehicular activities. The personnel, replenishment of supplies, delivery of experimental equipment, and return of data would be dependent initially upon ferrying spacecraft of the Apollo command-service module type.

The configuration selected for study by the Task Group was a 12-man module capable of being expanded into a composite space station to accommodate as many as fifty men as the need arose.[4] It was to be a centralized station designed for carrying out research, involving inter-disciplinary experiments and as an earth-orbital operation and maintenance depot for unmanned satellites. The life of the space station was set at ten years, if periodically maintained and resupplied. Essential, in the designing of this modular centralized station, was the extent to which it would contribute significantly to the reduction of the cost of future space missions.

Considerable momentum developed in the Apollo Applications Program for implementing the Skylab Project and launching a space station as soon as possible after the cessation of the Apollo moon-landing missions.[5] This, it was hoped, would fill the void and give renewed life to the nation's sagging space programs. Such a revitalization was needed if the potentials of space exploration and exploitation were to be realized in the coming decades.

Attention centered, prior to 1969, on the utilization of a modified spent Saturn 1B rocket for the nation's first earth-orbital space station.[6] The solar astronomy observatory was to be launched separately by a Saturn 1B carrier rocket, after which the two spacecraft would be docked while in earth orbit. Later, it was decided that the

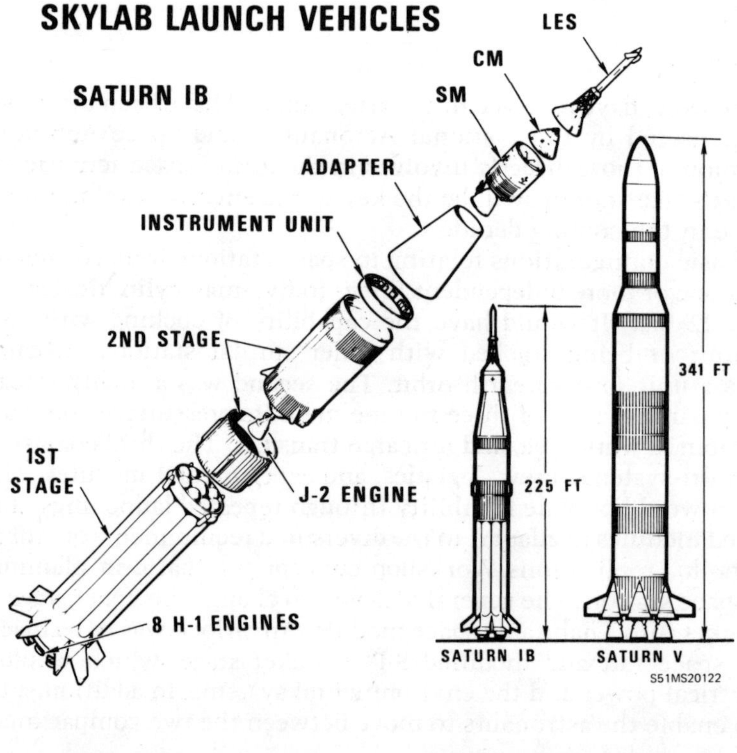

Skylab Launch Vehicle Diagram (Source: NASA - Photo No. 71-H-1104)

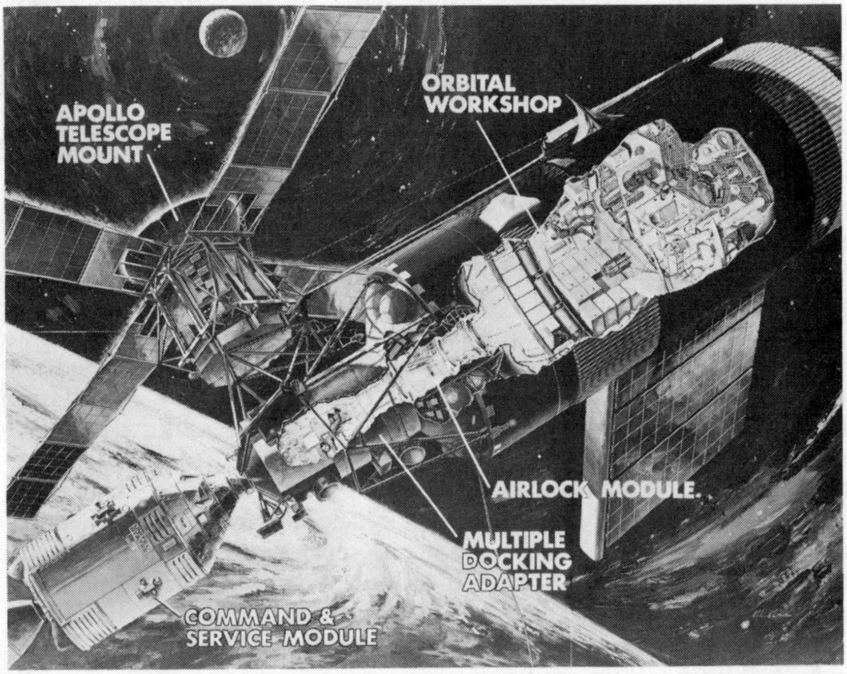

The Skylab Space Station (Source: NASA - Photo No. S-74-5148)

flight hardware, launch facilities requirements, and space flight operations would be simplified if the space station were assembled and outfitted on earth and then launched as a unit into an earth orbit by a Saturn V carrier rocket. In addition, this procedure would make for operation simplicity, greater safety, increased flexibility, reduction in the number of lift-offs, and increased payload capacity, all of which would result in important savings.

The simplified method consisted of two launches, the first utilizing a Saturn V launch vehicle would place the unmanned space station and the Apollo Telescope Mount in an earth orbit approximately 225 miles high an inclined 50 degrees to the equator. The earth orbit selected for its extended mission would place the space station over most of the heavily populated areas of the world. The second, a day later, and using a Saturn 1B launch vehicle, would place the Apollo command-service module, carrying three astronauts, in the same earth orbit. The two spacecraft would then dock and the astronauts would transfer to the space station and activate its systems. It was contemplated that they would remain aboard the space station for a period of twenty-eight days, which would be a record-making period. If all went well, there would be two revisits by three-man crews at three-month intervals.

The Experimental Workshop (Skylab)

The Skylab was a converted third stage of a Saturn V launch rocket. Inclusive of the Apollo command-service module, it measured 118 feet in length and weighed 100 tons. Its interior was equivalent to that of a three-bedroom house having 9,550 cubic feet. An important component of the Skylab was the Apollo Telescope Mount (ATM) with six telescopic cameras for observing the celestial bodies. This octagonally shaped 24,656-pound astronomical laboratory, with its windmill solar panels, was mounted on the exterior of the workshop and was operated from within. *It was the first manned astronomical laboratory, equipped with telescopic instruments, designed to study and record data relating to the sun and other celestial bodies from an earth orbit beyond the earth's atmosphere.*

The amenities provided for its occupants were far in excess of those that were available for the astronauts of the Apollo missions. The greatest advantage lay in the space available for the astronauts to live and work during their orbital tours of duty. Separate trays and individual heaters were provided for the preparation and heating of a wide variety of frozen consumables, as well as specially designed squirt-water bottles from which half an ounce of liquid could be sipped at one time. Running water was available for bathing by means of a handwasher which released water into a washcloth squeezer by pressing a button. This device was equipped also with a washcloth and a waste-water bag. Included in the Skylab's installations was a zero-gravity shower with a specially designed enclosure, making it posible for the astronauts to take periodic baths while in space. In addition, individual sleeping compartments were designed to permit them to sleep in a suspended manner which was not uncomfortable because of the lack of gravity. The Skylab had enough food and drink to sustain the members of the separate crews for the duration of the entire mission, being stocked with 720 gallons of drinking water and some twenty-four hundred pounds of food. There were also stowed aboard enough medical supplies and scientific equipment for the whole mission.

The overall purposes of the Skylab mission were to carry out scientific investigations in earth orbit, develop methods for evaluating earth environments from space, and acquire more detailed knowledge of man's capability to live and work under the conditions of weightlessness. The major scientific efforts were to be dependent upon the use of the solar astronomy observatory, with its six telescopes, for observing the sun from various parts of the spectrum. Further observations were to be made in connection with meteorology, communications, material processing and earth's resources. Multispectral photography was to be employed for gathering data relating to oceanography, water management, forestry, agriculture, and ecology.

Earth Resources Surveys, employing specially designed sensoring devices were to constitute an important phase of Skylab's scientific objectives. The earth orbit, in which the Skylab was to circle the earth every 93 minutes would permit the astronauts to observe a swath of the earth's surface stretching from 50 degrees North to 60 degrees South of the equator, an area containing 75 per cent of the surface of the earth, including the United States (except Alaska), and 90 per cent of the population of the world.

On the agenda were experiments having to do with material processing in space by means of electric furnaces and electronic beams. It was believed that the knowledge gained from these experiments would eventually lead to space manufacturing of strong welded metals, near-perfect ball-bearings, precision optical lenses, pure vaccines, and other products not capable of being produced under earth gravitational conditions.

Among the important experiments to be conducted during the overall mission were those pertaining to the health and welfare of the astronauts. They were designed to determine their ability to withstand zero gravity conditions and record the amounts of food consumed and amounts of liquids and solid wastes excreted by them daily. They were to take samples of their blood and record their blood pressures regularly. For three successive nights they were to alternate in wearing an electronic cap to record their brain-wave patterns while sleeping. Part of the space station's equipment was to consist of light-weight exercising machines to determine the capacities of the astronauts for performing strenuous exercises under conditions of weightlessness.

On January 18, 1972, the National Aeronautics and Space Administration announced the members of the three crews of the Skylab mission. The members of the Skylab-2 crew, the initial group of astronauts, who were to occupy the Skylab for 28 days, were Captain Charles Conrad, Jr., a veteran of three previous space flights,[7] commander, and Commander Paul J. Weitz and Dr. Joseph P. Kerwin, a physician astronaut. The members of the Skylab-3 crew who were to stay aboard the space station for 56 days after the departure of the first crew, were Captain Alan L. Bean,[8] commander, and Major Jack R. Lousma and Dr. Own K. Garriott, a scientist-astronaut, specializing in electrical engineering and atmospheric physics. Following the departure of Skylab-3 crew, Skylab-4 crew, under the command of Lieutenant-Colonel Gerald P. Carr, accompanied by Lieutenant-Colonel William R. Pogue and Dr. Edward G. Gibson, a scientist-astronaut, specializing in plasma physics, were to man the workshop and remain there for another 56 days.

On April 16, 1973, the Skylab was transported to its launch pad, mounted on top of a Saturn V carrier rocket, for a four-week checkout prior to the scheduled lift-off at Cape Kennedy on May 15. A large tracked platform carried the spacecraft and the carrier rocket 3.5 miles from the assembly building to the launching pad taking six hours to complete the transfer. The initial crew members, who had been in quarantine since April 24, were moved to quarters five miles from the launching pad from where they could view the launching. About 8,700 feet from the Skylab launching site stood the Saturn 1B carrier rocket, with the Apollo command-service module on top ready to carry the astronauts into space the following day should the launching of Skylab prove successful.

On May 14, at 1:30 p.m. (EDT) unmanned Skylab was hurled into space in a flawless lift-off by the 33-story Saturn V carrier vehicle and ten minutes later was positioned in a 272 mile earth orbit. Mission Control announced before it had completed the first orbit, that two of the six solar panels, damaged or jammed after take-off, had failed to deploy fully resulting in the space station losing about half of its electrical power.

On May 15, Skylab developed a heat problem when its interior temperatures rose to an average 110 degrees Fahrenheit, and there was little hope of launching the astronauts on schedule. It was determined that the problem was caused by the loss of an aluminum micrometeorite shield, ripped from the side of the space station 63 seconds after lift-off at 40,000 feet, removing the insulation cover from a portion of the space station's fuselage. The onboard air conditioning system could not be activated because it would deplete the highly restricted electric power supply. But the engineers succeeded in stabilizing the temperatures, between 90 and 105 degrees, by turning the exposed portion of the space station away from the direct rays of the sun.

The engineers had three options with respect to the disabled space station: to send the Skylab-2 crew into space to inspect it, to send them on a mission to attach a thermal blanket to it, or to abandon it and start work on the back-up space station.

After the basic problem had been defined, the engineers decided to construct an insulating canopy of paper-thin plastic material (Mylar) capable of reflecting and blocking the sun's heat. It was to be spread over the damaged exterior of the space

The Skylab with Deployed Parasol Solar Shield. This photograph was taken from Skylab 2 ferry command-service module during the final flyaround of the space station. (Source: NASA - Photo No. 73-H-589)

station by the astronauts when they arrived at its location. Three methods of installing a sun shield were studied: placing a window-type shield over the exposed part during a spacewalk by an astronaut, placing a shield over the exposed part by an astronaut standing in the hatch of the Apollo ferry spacecraft, and deploying a shield out of the airlock hatch of Skylab with the astronauts remaining in the hatchway.

Skylab 2 Mission (The Initial Manned Skylab)

After the situation had been studied for about eight hours, the lift-off of Skylab 2 flight scheduled for May 15, was postponed until May 20, in order to provide time for the engineers to devise a way of salvaging the crippled workshop. On May 17, the launching was postponed again, this time until May 25, which proved to be the last delay. The second five-day delay was to provide time for training the astronauts for the salvage operation with a minimum of risk.

On May 25, at 9:00 a.m., the Saturn 1B, after being delayed twice for a total of ten days, blasted off its pad at Cape Kennedy and within ten minutes placed the Apollo ferry spacecraft with its crew Captain Charles Conrad, Jr., Commander Paul J. Weitz and Dr. Joseph P. Kerwin in earth orbit. After a pursuit of seven and a half hours, the astronauts overtook the crippled Skylab at 4:30 p.m. and circled it to assess the exterior damage. They found that one of the power panels had been sheared off completely and the other jammed.

About two and a half hours after overtaking Skylab, they docked successfully and then undocked at 7:30 p.m. to attempt freeing the jammed solar-power panel, which, on first inspection, appeared to be a relatively easy task. Astronaut Weitz leaned out of the hatch of the Apollo spacecraft, with Astronaut Kerwin holding him by the knees, and attempted to carry out the repair task using a ten-foot shepherd's crook and a pole with a pruning fork attached at one end. Meantime, Astronaut Conrad maneuvered the Apollo spacecraft close to Skylab. Their efforts to free the solar-power panel were unsuccessful, largely because of inappropriate tools for the task. *This was man's first attempt to carry out a repair operation outside a spacecraft in space.* It was not a true space walk because the astronaut did not leave the Apollo command module.

In attempting to redock the Apollo command module with Skylab, the astronauts encountered a malfunction in the docking mechanism. After executing nine unsuccessful docking maneuvers, Captain Conrad attempted to remedy the situation in an unorthodox manner. The air was removed from the cabin of the Apollo spacecraft and the astronauts, clothed in their bulky spacesuits and bubble helmets, manually repaired the docking mechanism and achieved a firm docking with Skylab at 11:50 p.m.

The astronauts spent the first two nights in space in the Apollo ferry spacecraft because of the intense heat of the interior of Skylab. On May 26 at 12:45 p.m., they entered it wearing gas masks and carrying sniffer cannisters as a precaution against the possibility of the presence of noxious gases. Following a quick inspection, they proceeded to deploy the orange and silver sunshade over the damaged portion of Skylab's fuselage taking until 9:00 p.m. to do it. This was accomplished by attaching the collapsed sunshade to a five-section 21-foot pole, and passing it through the narrow airlock in the wall of the spacecraft's fuselage. They then released the mechanism to deploy it and at the same time, withdrew the pole to position the 22′ x 24′ sheet above the fuselage. They did not achieve full deployment of the sunshade but it provided sufficient coverage of the damaged portion of the fuselage to reduce gradually the interior temperature to between eighty and ninety degrees Fahrenheit. This was still hotter than desired but the astronauts were able to occupy Skylab and carried out part of their assignments. At 2:25 p.m. on May 27, Astronaut Weitz entered the laboratory and found it to be hot but otherwise in good condition. He was joined later by Astronaut Conrad and they proceeded to activate the space station.

The first attempt to free the jammed solar power panel was unsuccessful, and the subsequent power situation made imperative another attempt to assure the survival of the mission. The failure of some of the batteries and the possible failure of others, gave cause for concern, especially because Skylab's power was being supplemented by the power of the Apollo ferry spacecraft of limited duration. When the four operative solar-power panels were turned away from the sun, forcing Skylab to reiy on its 17 batteries, (one inoperative), four of them clicked off. The space station passed from daylight to night and back again during each 93-minute orbit. While in the sunlight the solar panels were operative, whereas they were inoperative while in the dark. After the solar panels resumed operation, a fifth battery failed, and the batteries had to be reactivated by signals from ground control. Under the circumstances, the earth photographic surveys planned for May 31, were cancelled and on June 7, another battery began to lose power making the situation more precarious.

In spite of the unfavorable conditions prevailing in the space station, the astronauts proceeded to initiate as many scientific experiments and make observations as the limitations of power and high temperatures would allow. On May 29, they turned on the array of telescopes to observe the X-ray emissions and examine the structural aspects of the sun's atmosphere. *This marked the beginning of a new era in astronomy involving observations from above the earth's atmosphere.*

The following day, they undertook a survey of the earth in spite of continuing possibilities of having to further limit their scientific activities. They photographed Great Salt Lake and Lake Powell in Utah, gypsum deposits of White Sands, New Mexico, and soil salinity of Rio Grande Valley. In addition, they attempted to take mapping pictures of Central America and resources survey pictures of South America, but were hindered in these activities by cloud coverages. The earth resources survey was the third and final major experiment undertaken that week. May 29 was the first day since being in space that they had come near meeting their planned assignments.

Mission Control decided that Captain Conrad and Astronaut Kerwin should undertake another space walk and attempt to free the jammed solar panel. Astronaut Weitz was to remain in Skylab as the communication link with Mission Control. On June 7 at 11:23 a.m., the hatch in the airlock compartment was opened just before passing out of radio contact and Captain Conrad took his position outside the spacecraft, attached to it by a 60-foot umbilical cord. Astronaut Kerwin passed out sections of tubing to be used as a make-shift handrail to assist in reaching the jammed solar panel 25 feet away. He then left the space station to assist Astronaut Conrad in freeing the solar panel. They felt that they had a fifty-fifty chance of freeing it and after about five hours, using a bolt-cutter attached to a 25-foot pole, they succeeded in cutting through the aluminum strip restraining the panel. Immediately the panel deployed and the power surged upward when the sun's rays activated its numerous cells. While outside the spacecraft, they repaired a cover on one of the solar telescopes, replaced a film cassette in another, checked a thruster, and inspected the sunshade.

With Skylab's power difficulties partially overcome, the astronauts turned their attention more fully to their research assignments. They began to scan the earth with the cameras focusing on corn fields of Nebraska, strip-mining in Kentucky, and flood conditions in the Mississippi Valley. They also took pictures of the midwestern and northwestern parts of the United States on June 9, and on June 11 surveyed a storm over Texas and Oklahoma. An experiment having to do with molten metals under conditions of weightlessness was carried out on June 12. They were now more optimistic about completing their 28-day assignment and continuing the overall mission of Skylab. There remained still the problem of the malfunctioning coolant loop which itself was not a direct threat to the mission because its temperature had been stabilized at 35 degrees (Fahrenheit). It was thought that the cause of this problem was a frozen valve which could in time be remedied effectively.

One of the exciting experiences of the Skylab-2 crew was photographing the solar flare which occurred suddenly on June 15, and continued for about ten or fifteen minutes. Astronaut Weitz focused the telescopic cameras on it within minutes after it started and continued to photograph it as it rose in intensity and declined. It began as a pinpoint of light ten times brighter than the surrounding sunlight and built up in intensity, spiraling outward and ending in a tremendous explosion of a huge mass of solar gases. It has been estimated that such solar phenomena produce as much energy as man has consumed over several decades.

On June 18, Skylab-2 astronauts surpassed the space endurance record for a single mission set by Russian cosmonauts in June, 1971.[9] *Captain Charles Conrad, Jr. himself set a record for the total individual flight time, having logged more than forty-four days in space during four space flights.* These records were not expected to last in view of the impending long missions planned for Skylab-3 and Skylab-4 crews.

On June 19 at 6:55 Astronauts Conrad and Weitz began the fourth space walk to recover six cannisters of films from the solar telescopes. While in space, Astronaut Conrad revived a dead battery by striking it with a hammer.

In preparation for their departure from Skylab and return journey to earth, the Skylab-2 astronauts spent their last full day in space putting the orbital workshop in order for the occupancy of the Skylab-3 astronauts. On Friday, June 22, they undocked and, after executing several maneuvers, were earthward bound. They splashed down in a calm Pacific Ocean at 9:30 a.m., on June 22, on target and on schedule about six and a half miles from the recovery ship *USS Ticonderoga* and 840 miles southwest of San Diego. Unlike previous Apollo splashdowns, they remained inside the Apollo capsule, which was hoisted aboard the recovery ship within forty minutes after splashdown. It was not certain whether they would be able to leave the capsule without assistance or would have to be carried out on stretchers, because they had been exposed so long to conditions of zero-gravity. However, they emerged from the capsule under their own power and, although somewhat unsteady, walked across the deck to the medical laboratory to undergo medical examinations. They received a message from President Nixon welcoming them back and inviting them to visit with him Sunday morning at the Western White House at San Clemente where they would meet Soviet Communist Party Leader Leonid I. Brezhnev.

Their stay in space appeared to have had no permanent ill effects, although the return to the earth's gravitational environment did cause some dizziness and nausea. These symptoms were temporary and within a matter of hours they began to regain their normal physical state and were able to move freely with few inhibitions. Astronaut Kerwin took longer to recover from the discomfortures than did Captain Conrad and Astronaut Weitz. Captain Conrad showed the least effects and experienced the speediest recovery. In general, the evidence supported the contention that man would be able to adjust to long period in space. In spite of the numerous difficulties encountered by the Skylab-2 astronauts and the adverse conditions under which they had to live and work in space, they had succeeded in accomplishing about eighty per cent of their assigned scientific experiments.

The Skylab-2 crew succeeded in operating the space station's eight solar telescopes for 81 out of a scheduled 100 hours, observing ultraviolet and X-ray solar emissions, took 30,000 photographs and covered 60 per cent of their assigned tasks with respect to earth resources surveys, having taken 14,000 pictures over 31 states and photographed hurricane Ava. *In addition, they established several space records, having successfully salvaged the largest and heaviest spacecraft ever put into earth orbit; completed the largest number of revolutions of the earth (391); flew the greatest number of space miles (11.5 million); spent the longest time in space (28 days and 50 minutes); achieved the longest time in space for an individual (Astronaut Conrad) 48 days, 1 hour and 8 minutes; accomplished the first successful repair operation outside an orbiting spacecraft by freeing the jammed solar power panel; threw the first darts under weightless conditions; and performed for the first time gymnastics and tumbles by running around inside an orbiting spacecraft.*

Skylab 3 Mission

The Saturn 1B carrier rocket and the Apollo ferry spacecraft, for the Skylab-3 mission, were moved to the launching pad on June 11. The flight was originally scheduled for August 8, 1973, but because of the power problem of Skylab, was advanced to July 27. On Saturday, July 28, 1973, at 7:11 a.m. the Skylab-3 mission was initiated, when the Apollo spacecraft with Captain Alan L. Bean[10] in command and accompanied by Major Jack R. Lousma and Dr. Owen K. Garriott, a scientist, was hurled into an earth orbit within ten minutes after lift-off on a 59½-day orbital flight. The spacecraft went into an initial earth orbit with an apogee of 140 miles and a perigee of 94 miles. At 3:43 p.m., after circling the earth five times, it made a rendezvous and docked with Skylab following a chase of 2,500 miles. On the first maneuver to overtake Skylab, one of the four pairs of thruster rockets developed a leak, but this was not serious enough to interfere with docking with the space station.

The objectives of the Skylab-3 mission were similar to those of Skylab-2, but were more extensive in observations and scientific experiments. The astronauts were to conduct experiments involving observations of the sun through high resolution telescopic cameras, surveying the earth's resources, using multispectral cameras and sensing devices, and examining themselves for physiological changes resulting from living and working under zero conditions of gravity. It was anticipated that the astronauts would utilize the solar telescopes three times more than their predecessors.

They took along with them a menagerie consisting of mice, fish, gnats, spiders, and other insects. In addition, they took rice seeds and plants. These were to be used in a series of biological experiments to determine the impact of zero gravity on the biological rhythm of living organisms. The minnows and the cross spiders (Arabella and Anita) were to be taken into the space laboratory for extended observations. The spiders, which received wide acclaim, were to be observed as to their ability to spin webs in space and the eggs of the minnows, if they hatched, would provide an opportunity for observing the behavior of space born fish and comparing it to that of adult fish not born in space. The study of the gnats should indicate in what manner their daily physiological rhythm (biological time-clock) was affected by zero gravity. The body temperatures and the effects of weightlessness on the mice were to be monitored automatically.

The astronauts experienced considerable nausea during the first few days onboard Skylab. This retarded their activities and necessitated delay of the scheduled space walk to deploy the new sunshade and replace films in the telescopic cameras. It was re-scheduled for August 1, and after several postponements finally took place on August 6.

On July 30, the astronauts reported that the nitrogen-oxygen air inside the space station had become quite humid. This was due to a leak in the space station's system for withdrawing moisture from the air. On August 2, they discovered leaking steering rockets on the Apollo spacecraft which was to ferry them to earth on completion of their mission. When a second set of maneuvering rockets developed malfunctions and had to be shut down, Mission Control inaugurated a round-the-clock program to prepare an Apollo spacecraft and a Saturn 1B carrier rocket for a possible emergency rescue mission.[11] There was considerable concern about the capability of the Apollo ferry spacecraft docked with Skylab to return the astronauts safely to earth. Astronauts Vance P. Brand and Donald L. Lind were assigned to fly the rescue mission should the need arise. It would not be possible to launch the rescue spacecraft before

September 10, because of the time required to prepare it for such a mission. The Apollo ferry spacecraft had to be converted from a two-man to a three-man space vehicle.

The space officials were confronted with three options with respect to the potential emergency: continue the activities of the Skylab 3 astronauts under day-to-day observations, launch the rescue mission between September 10 and 15, or require the immediate termination of the mission and bring the astronauts back in their Apollo ferry spacecraft. The engineers decided to follow the first option because tests indicated that the Apollo ferry spacecraft would continue to perform until the end of the mission and be capable of returning the astronauts safely to earth. It was felt that procedures could be developed to adequately maneuver the spacecraft to a safe landing without the full complement of steering rockets. They had not identified, as of that time, the source of the malfunction in connection with them.

August 4 was a difficult day for the astronauts. They were awakened during the night because of a short circuit in the telescopic unit causing the failure of a television camera to monitor the solar atmosphere. At another time during the night, there was concern that a more serious malfunction had caused a loss of pressure in the helium tank, but it turned out that this was because of incorrect data processing. There was also a false fire alarm. In addition, one of the cameras had to be jettisoned because it would not retract through an airlock. There was still the problem of the leaky steering rockets on the Apollo ferry spacecraft.

On the night of April 5, another problem arose in connection with the air conditioning system. There was no immediate danger to the astronauts, but if it were not remedied, it might bring about the early termination of the mission. It was estimated that there was a sixteen-day supply of liquid coolant in the primary system and a sixty-day supply in the secondary system.

The repeatedly postponed space walk was carried out on August 6, for the purpose of deploying a new paper-thin aluminized Mylar sunshade over the old one to provide additional protection to the damaged roof of Skylab from the rays of the sun, replacing the films in the telescopic cameras, and inspecting the leaky rocket units of the Apollo ferry spacecraft. This space walk established an endurance record of 6 hours and 31 minutes and was carried out by Major Jack Lousma and Dr. Owen Garriott, with Captain Alan Bean monitoring the controls inside Skylab.

Astronaut Garriott stood in the hatchway of Skylab and passed two 55-foot poles, which he assembled out of five-foot sections to Astronaut Lousma, standing at the base of the telescope unit. On receiving the poles, Astronaut Lousma attached them to the base of the telescope unit to form a "V" extending over the damaged part of Skylab. Pulleys had been attached at the far ends of the poles with ropes threaded through them. The folded 12 x 24-foot sunshade was then hooked to the rope and pulled into position. Some difficulties were encountered in maneuvering the sunshade into position because at first it would not unfold fully. Before completing the space walk Major Lousma inspected the leaky steering rockets and reloaded the telescopic cameras. He also positioned a set of panels for detecting the impact of micrometeorites and measuring their size, mass, and velocity.

On August 7, following a period of lingering motion sickness, nights of threatening mishaps, and extravehicular activities, the astronauts were able to direct their attention to the scientific objectives of the mission. Dr. Garriott operated the onboard telescopes for the first time and spent three hours focusing them on the outer atmosphere of the sun where small flares had been observed. The temperature inside

Skylab had dropped to about eighty degrees largely because of the installation of the new sunshade and in general living and working conditions were much improved. The only new malfunction of any significance was the onboard teleprinter, but this was corrected by placing a space printing mechanism in the machine. These improved conditions did not lead Mission Control to discontinue the preparations for a possible rescue operation.

The three astronauts found that they had become more adapted to the conditions in Skylab and were able to perform their various assignments with more dexterity. In view of this, they requested that they be assigned more work to keep them occupied because they were completing their tasks ahead of schedule. Mission Control, in response increased their workload by 11 man-hours a day. It was also found that five to six hours of sleep were sufficient under space conditions.

Much of the astronauts' time was spent in observing the sun which was particularly turbulent throughout the mission.[12] On August 10, telescopic observations were made of a truly spectacular solar disturbance. It was an eruption of gaseous materials in the western rim of the sun and in a direction away from the earth.[13] It was estimated that the mass dispersed was greater than the mass of the earth.

On August 21, the astronauts photographed another solar phenomenon. This time it was a huge bubble in the outer atmosphere of the sun. It was thought to be the most significant solar observation of the mission and was believed to have been caused by an explosion on the far side of the moon. It was assumed to be three quarters the size of the sun.

On September 6, they observed and photographed the aftermath of a solar eruption estimated to be one hundred times as powerful as the San Francisco earthquake of 1906. They discovered a half-elliptical blob in the corona of the sun thought to be the final release of energy of a huge solar flare. Again, on September 7, they photographed another solar explosion. This time the mass appeared to be ten times that of the earth the largest and most intense of the year, and was believed to be equal to the force of one hundred million atomic bombs. Following the peak of this solar disturbance, there occurred a period of 20 minutes when long distance short-wave radio communications on earth faded out completely.

While they were engaged in making observations of the sun, they were also conducting surveys of the earth's resources. On August 11, they collected information relating to storms, forests and mineral deposits, covering a sweep of 63,000 miles of the United States and Latin America. Another survey was made on the following day, when the southwestern part of the United States was surveyed with particular attention given to California, Nevada, New Mexico and Texas.

On August 31, they watched the tropical storm Christine from the time it started as a depression and developed into a full-size storm. The storm was 900 miles east of Barbados and moved westward at 14 miles an hour with the strongest sustained winds reaching fifty to sixty miles an hour.

Another space walk was undertaken on August 24, extending over four and one-half hours. Astronauts Lousma and Garriott loaded unexposed films into the telescopic cameras and installed a new set of gyroscopes. Major Lousma plugged in four connections on a 23-foot gyroscope cable while Dr. Garriott placed films in the cameras, adjusted two doors on the telescopes and retrieved a space-dust collection experiment. Meantime, Captain Bean was inside Skylab monitoring the controls to maintain stability while the gyroscopes were being replaced.

Astronaut Garriott Performing Extravehicular Activity Near the Apollo Telescope Mount of the Skylab Space Station Cluster in Earth Orbit, Skylab 3 (Source: NASA - Photo No. 73-RC-750)

By the time the mission had reached its half-way mark, the Skylab-3 crew had surpassed the space endurance record. As of August 27, the members of the crew had spent a month in space and were at the halfway point, having circled the earth 429 times, traveling more than twelve million miles. On completion of the Skylab-3 mission, Captain Bean would have accumulated slightly more than 69 days and 15 hours in space. It was thought, at the time, that the record, if achieved, would stand for a relatively long time, because the Skylab-4 crew was composed of space rookies and the United States had not scheduled any long-duration missions to take place before the end of the decade.[14]

An interesting experiment was conducted on September 10, when crystals were melted in a special furnace. Parts of two crystals were transformed into a molten blob and then allowed to cool and reform into crystals. The results of this experiment may eventually lead to the production of pure metals with strengths and durability increased a thousandfold.

The third and final extravehicular activities were successfully carried out on September 22 by Captain Bean and Dr. Garriott. They spent 2 hours and 41 minutes outside Skylab in recovering six cannisters of exposed films. Captain Bean accomplished the replacement of the films after climbing a seven-rung ladder and passing the retrieved cassettes to Dr. Garriott standing in the open hatchway. Major Lousma remained at the controls inside Skylab.

September 25 was the last day spent by the crew in Skylab. After putting it in order for the visit by the Skylab-4 crew, scheduled for November, they sealed it and entered the Apollo ferry spacecraft about 11:30 a.m. They realized that if they were to return to earth safely they would have to employ techniques never before tried in space. These complex procedures had been formulated by Mission Control after long and painstaking deliberations.

After closing the hatchways, they undocked from Skylab, allowing the Apollo spacecraft to drop out of its existing earth orbit and at 6:04 p.m. plunged through the earth's atmosphere. The crippled spacecraft with the astronauts inside, landed on a choppy Pacific Ocean on the same day at 6:20 p.m. about two hundred and thirty miles southwest of San Diego and six miles from the recovery ship. Within an hour, the space capsule, with the astronauts still inside, was hoisted aboard the USS *New Orleans,* where they received a message of congratulations from President Nixon, who had received a congratulatory message from President Nikolai V. Podgorny of the Soviet Union.

The successful splashdown brought to an end a space mission that had lasted for 59 days, 11 hours, 9 minutes and 4 seconds and had covered 24.4 million miles. Its scientific achievements were outstanding. The astronauts brought back with them, 77,600 solar photographs, 16,800 pictures of the earth and data requiring 18 miles of magnetic tape to record them. They had completed 305 hours of viewing time with the solar telescopes as compared with 200 planned hours and had carried out 39 successful passes in earth-resources surveys compared with 26 planned sweeps.

The information gathered by the Skylab-3 crew should provide a better understanding of the sun and its turbulent outer atmosphere, the temperatures and moisture contents of tropical storms, location, quantity, and quality of the earth's resources, and the characteristics of pure metals produced under conditions of weighlessness. The biological studies should provide insight as to the biological rhythms of living organisms. In addition, a greater appreciation of man's capabilities to live and work under conditions of weightlessness and of his place in the universe should be derived from the vast accumulation of data.

Skylab-4 Mission

Following the difficulties encountered with the space station, there was some question as to when the Skylab-4 Mission should be undertaken. It was tentatively scheduled to be initiated on November 10, 1973.

On October 23, during a fueling operation, the domes of two partly filled fuel tanks of the booster stage of the Saturn 1B carrier rocket collapsed. It was determined that this had been caused by a plastic cover placed over the tanks to protect them from the rain. It had prevented air from entering the tanks as the kerosene was withdrawn and a vacuum created in the upper part of them with the result that the outside air caused them to buckle. It was decided that by pressurizing them the domes would resume their former shapes. If this did not work, the mission would have to be postponed until December to allow time for bringing another Saturn 1B to the launching pad. The tanks were pressurized and the domes resumed their proper form without suffering any structural damage. This mishap did not necessitate a postponement of the scheduled date for launching the Skylab-4 spacecraft.

During a routine check of the Saturn 1B rocket on November 6, there were discovered on its eight fins 14 long hairline cracks, caused by the long exposure of the rocket to the salty air.[15] The fins had to be replaced and the launching of Skylab-4 was delayed five days, until November 15. Each fin weighed 473 pounds and was 16 feet high and spread out to nine feet at its base. The replacement of the eight fins was a time-consuming task. Delays and discovery of other cracks slowed down the repair work and on November 12 a further delay of 24 hours was ordered by Mission Control.

Finally on November 16, 1973, at 9:01 a.m. the Skylab-4 spacecraft was hurled into space, carrying Lieutenant Colonel Gerald P. Carr, commander, accompanied by Lieutenant Colonel William R. Pogue and Dr. Edward G. Gibson, a scientist-astronaut, all making their first space flight. Within eight hours after lift-off and five rocket firings, the astronauts accomplished a rendezvous with the space station 270 miles above the earth and, after two unsuccessful attempts, docked with it at 5:02 a.m. The mission had been extended from 56 to 84 days, following studies of the capabilities of the space station, the reserves of oxygen fuel, and medical tests of the astronauts of the previous Skylab missions.

The objectives of the mission were to observe the Comet Kohoutek with telescopes and special cameras before, during, and after its rendezvous with the sun; study the outer atmosphere of the sun, giving attention to the bright spots and coronal holes; photograph the earth's resources; test their physiological reactions to the extended exposure to zero gravity conditions, study the effects of near vacuum on the breeding of gypsy moths and their life-cycle patterns as a basis for possible development of sterile male moths.

For the first time in the Skylab mission, the astronauts of Skylab-4 were reprimanded publicly for attempting to conceal medical information, relating to Colonel Pogue's nausea on the first day of their occupancy of Skylab. His sickness was reported but not the fact that he had vomited. In fact they had discussed dumping the vomit but did not do so. The situation came to light when a recording was reviewed by Mission Control of a conversation between the astronauts who were not aware apparently that the recorder was operating at the time.

This incident in retrospect seems to have been the forerunner of other disagreements and inadequate communications between the astronauts and the ground controllers, persisting throughout the greater part of the mission and surfacing from time to time. An overly ambitious pre-mission scheduling of work may have been largely responsible for much of the disagreement and misunderstanding. Upon occupying the space station, the astronauts had numerous housekeeping tasks to perform, including sorting out and transferring several hundred pounds of items, consisting of food, films, magnetic tapes and other equipment from the Apollo ferry spacecraft. Procedural errors were made also in activating the space station. On one occasion, the wrong valve was opened and an antiseptic solution used for sterilizing the water was dumped into space. In another instance, filters were not installed in an array of cameras resulting in the loss of data in connection with an earth resources survey. Criticism was leveled also at the pre-mission planners for including certain medical tests for which the astronauts were not adequately trained.

In time, these problems were largely eliminated and better relationships existed between the astronauts and the ground controllers. This was especially true after free and open discussions of the problems and issues between the parties evolved. It may have been that too much publicity was given to the reactions of the astronauts and their attitudes toward their work and comparing their performance with that of the Skylab-3 crew. One official came to the defense of the astronauts, with respect to the published reports that they had made more mistakes as a crew than the other crews of the Skylab Missions.[16] He insisted that they had performed well as demonstrated by their accomplishments up to that time.

On November 22, Colonel Pogue and Dr. Gibson spent 6 hours, 34 minutes and 35 seconds outside Skylab loading films in the telescopic cameras, repairing an antenna, and deploying a series of experiments on the hull of the space station. Colonel Carr remained inside the space station relaying instructions. This space walk was 3 minutes and 25 seconds longer than that performed by Astronauts Lousma and Garriott on August 6, as members of the Skylab-3 team.

A gyroscope on the Skylab failed on November 23, leaving only two to stabilize the space station, raising questions of what the impact would be on the duration of the mission should another fail. Even though the space station could be kept in operation with only two gyroscopes, this would necessitate greater fuel consumption by the maneuvering jets. It would also make it more difficult to position the space station for the precise aiming of the cameras and sensing devices at their respective targets and in performing onboard experiments.[17]

On November 27, Mission Control restricted further maneuvers of the Skylab because of the unexplained high consumption of gas by the jet thrusters used in stabilizing the space station. It was discovered that the jet thrusters were using six times as much fuel as planned. This restriction applied particularly to those maneuvers relating to the earth resources surveys and observations of the Comet Kohoutek. The following day it was determined that the high consumption of control gas was because incorrect data had been fed into the computer responsible for calculating control gas requirements of the jet thrusters; the restriction of maneuvers was then removed.

A considerable part of the astronauts' time was spent observing Comet Kohoutek. This was especially the case during the latter part of December 1973 and the early part of January 1974 covering the time prior to, during, and after its rendezvous with the sun on December 28. On December 5, they focused the powerful telescopes and cameras on the comet still some 120 million miles from the sun and traveling about 110,000 an hour. This was the first time that they had made an extensive study of the comet calculated to be fifteen to twenty miles wide and thought to consist of frozen dust and chemicals. They observed the comet on December 8, when it was some 75 million miles from the sun, traveling at 117,000 miles an hour and thought that it was developing a forked tail. During the previous two weeks, they had taken black and white pictures of it twice a day and had made frequent pictures of the comet with a special camera. By December 13, the tail of the comet was observable by the naked eye and appeared to be growing larger and brighter momentarily. On December 19, they collected extensive and valuable information on Comet Kohoutek, in spite of the trouble they encountered in focusing the solar telescopes on the rapidly moving object. They aimed four of the telescopic cameras, a hand-camera, and an ultra-violet light sensor at it. On the next day, they studied the composition of its core being vaporized as it drew nearer and nearer to the sun.

On Christmas Day, Colonel Carr and Colonel Pogue performed the second space walk of the mission. They carried three special cameras to be mounted on the outside of the space station to take pictures of the comet. This walk consumed seven hours, which surpassed the space walk taken by Colonel Pogue and Dr. Gibson on November 22. The primary purpose of this extravehicular excursion was to take pictures of the Comet Kohoutek which was then about 20 million miles from the sun. These pictures were costly in terms of control fuel consumption requiring about three thousand pounds of thruster gas, equal to more than ten per cent of the 26,000 pounds remaining for the rest of the mission. During the space walk, Dr. Gibson remained inside the spacecraft, using the jet thrusters to maneuver Skylab, because the two operating gyroscopes were unable to position the space station properly for the observations.

The comet made its rendezvous with the sun on December 28, at which time it came within 13 million miles of it. On the following day, Dr. Gibson and Colonel Carr walked in space, photographing the comet from outside the space station with special cameras. It appeared yellow and orange like a flame in space, and stretching toward the sun was a streak moving in a loop about its tail. This space walk provided an opportunity for a special photographic study of the comet, the results of which may provide clues as to the origin of the solar system.

While the observations and information gathering relating to Comet Kohoutek were in progress, the astronauts were also busy carrying out other assignments, especially making surveys of the earth's resources. On November 30, they conducted a photographic surveillance of a 6,900-mile sweep of coal miles in Illinois, Indiana, and Kentucky to determine their effects on the environment. This survey also took in the Atlantic coastline of Georgia and South Carolina for wildlife wetlands and parts of Brazil for information about thunder storms.

On January 8, they carried out a survey of Guatemala, looking for volcanoes as potential sources of energy. This photographic surveillance covered a sweep of 6,800 miles, stretching from the Pacific Ocean across Central America and Florida to the south of France. Again, on January 9, they were engaged in a search for geothermal spots which might prove to be sources of power. They employed cameras and heat sensoring instruments in their efforts to locate such spots in Nicaragua and Honduras, where the interior heat of the earth comes near the surface.

Investigations were made of the shifting desert sand dunes and the drifting ocean currents on January 12, during their twenty-fifth survey of the earth. The area was similar to that previously surveyed to provide a more thorough coverage of the United States, Mexico, and Central America. The dunes of Southern Saudi Arabia were photographed with hand cameras and later the cameras were focused on the Gulf Stream near Florida to study the changes as it approached land masses. On January 21, pictures were made of a Mississippi cotton growing area to obtain data to assist cotton farmers in obtaining better yields per acre. The sweep covered 8,000 miles stretching from Northwestern United States as far as South America. With a little more than a week left in the mission, the astronauts continued their earth resources surveys and on January 29, carried out another survey, this time of a 5,500-mile swath covering the United States and South America. On February 1 they made their final earth resources sweep of 8,500 miles.

The final extravehicular activities, lasting 5 hours and 19 minutes, were successfully accomplished on February 3. Astronauts Carr and Gibson went outside the space station to recover exposed films from the telescopic cameras, and Colonel Pogue remained inside to relay instructions to them. They also took pictures of the sun with an X-ray sensitive camera. Colonel Carr undertook the difficult task of removing a plate from the side of the space station to be taken back to earth for examination to determine the effect of nine months in space upon it.

On February 6, the space station was maneuvered into a higher earth orbit, so as to reduce the drag of the earth's gravitational force. This should extend its duration in space and make more probable revisits to it.[18] The astronauts left, in the sealed Skylab, a sample tray of food, films, filters, teleprint paper, and an electric cable, so that if it were revisited, these objects would provide some indications of the effects of prolonged exposure to zero gravity upon them.

After sealing Skylab and closing the hatches, they undocked from the space station on February 8, 1974, some six hundred miles east-northeast of Bermuda while in their 1,214th orbit of the earth and Skylab's 3,898th earth orbit. They followed the usual sequence of maneuvers for reentry and after jettisoning the service module, reentered the earth's atmosphere. At 11:17 a.m. (EDT) on the same day, the space capsule splashed upside down in the Pacific Ocean 176 miles southwest of San Diego and 3.4 miles from the recovery ship after 84 days, 1 hour and 16 seconds in space. The capsule was hoisted to the deck of the USS *New Orelans* and the astronauts were assisted from it, after a 20-minute medical examination, to a fork-lift to be carried by it across the ship's deck to the medical laboratory for further examinations. Although they were wobbly on their legs, they were in good health and cheerful. It was later reported that their physical condition was better than that of the previous Skylab crews, upon their return to earth. While on the recovery ship, they received a radio message from President Nixon, congratulating them on their splendid accomplishments in space.

Despite their misunderstandings and inadequate communications with the ground controllers, the astronauts performed well and completed a highly successful scientific mission. They observed the sun for 338 hours with an array of onboard-telescopic cameras, focused a second array of cameras and instruments on the earth, making 39 separate surveys and carried out many experiments, including making metal alloys and ultra-pure crystals for use in electronics. They even made photographs of laser beams originating on earth.[19] In addition, they repaired the air cooling system and an antenna, followed an intense schedule of physical exercises, performed new medical tests relating to the heart,[20] carried out four extravehicular excursions, made numerous observations of the Comet Kohoutek and returned with thousands of photographs of the sun, the comet and the earth's resources, together with a 100,000 feet of magnetic tape containing the results of scanning the earth's surface. They also established endurance records.

During the three manned Skylab missions, the nine astronauts accomplished a great deal in collecting data, consisting of 46,000 exposed frames of photographic films and some fifty miles of magnetic tape. It will take years to analyze fully the mass of data and to bring to light the wealth of information that should greatly expand our knowledge of the universe.

A fitting conclusion to this discussion of the Skylab Program was enacted at 2:50 a.m. (EST) on January 11, 1975 when the 83,500-pound second stage of the Saturn V carrier rocket responsible for orbiting the unmanned Skylab space station on May 14, 1973, plunged to a fiery destruction. The huge second stage rocket skipped off the earth's atmosphere over the Indian Ocean before ending its two years of tumbling out of control in orbiting the earth. Finally, it yielded to the force of the earth's gravity as it traveled from the North Atlantic to the Sahara Desert and reentered the earth's atmosphere and disintegrated. Few, if any, of its parts survived the intense frictional heat to reach the surface of the earth.[21] As for Skylab itself, it continues to circle the earth in a gravity gradient or slowly decaying orbit. It is anticipated that it will reenter the earth's atmosphere in 1981. There are at present no plans to revisit the space station.[22]

Shuttle Spacecraft

Essential to the ultimate success of the Skylab Project is the development of an integrated space transportation system based on a reusable space shuttle. The shuttle will not be operational until sometime after the termination of the Skylab missions, making it necessary, for a time at least, to rely on the Apollo command-service modules in combination with Saturn 1B launch vehicles to ferry the crews to and from the space station as well as to refurbish it and return data.

Four studies were begun during February 1969, sponsored by the Advanced Manned Mission Office (Space Shuttle Task Force) to determine the feasibility of a low-cost integrated space transportation system.[23] The studies provided three basic space shuttle configurations: a low-cost expendable launch vehicle in combination with an advanced reusable spacecraft, a one-and-a-half stage launch vehicle partially reusable with expendable tanks, and a fully reusable launch vehicle with recoverable booster rocket. The system adopted will have to be capable of lowering the cost significantly relative to the use of the Apollo spacecraft for ferrying men and supplies to the space station. In addition, it will have to be safe, flexible, versatile, and adaptable to more than purely logistical missions.

The studies set forth the following set of mission guidelines:[24]

1. Seek major reductions in the cost of recurring operations by planning to reuse systems with a minimum of refurbishment
2. Use a vertical launch systems with horizontal landings at fixed land sites
3. Design low-cost expendable elements for simple easy operational use

The Shuttle Spaceship of the Future, Artist Concept (Source: NASA - Photo No. 73-H-657)

A Space Shuttle Orbiter Approaching a Landing Field, Artist Concept. (Source: NASA - Photo No. 75-HC-257)

4. Keep ground support operations and equipment to a minimum for quick turn-around time
5. Provide a shirtsleeve environment with low "g" forces for passenger comfort
6. Allow for large integral cargo holds
7. Provide for passenger and cargo unloading through intra-vehicular transfer
8. Design vehicles large enough to accommodate a two-man crew and ten passengers.

The Space Task Group, reporting to the President in September 1969,[25] recommended that consideration be given to a reusable chemically fueled shuttle to operate between the surface of the earth and low earth-orbiting space stations, in the manner of a jet airliner, a chemically fueled reusable space tug for moving men and equipment between spacecraft in different earth orbits, usable also as a transfer tug between lunar oribt spacecraft and lunar bases on the moon's surface; and a reusable nuclear space shuttle for transporting men, spacecraft, and supplies between the low earth orbits and geosynchronous orbits and for other space activities.

In July 1969, the National Aeronautics and Space Administration designated the Office of Advanced Research and Technology as the agency responsible for the development of a base program for the implementation of the Space Shuttle Project. The general baselines for the study were the usability of major systems, elimination of the aerospike engine, because of its inability to provide adequate flexibility in the configuration of the shuttle spacecraft, and utilization of an engine with a thrust of approximately four hundred thousand pounds.[26]

Some space planners have predicted that the space shuttle will become the key component in ferrying men and materials to and from obiting space stations before the end of the century.[27] It may become also the major link in the transportation of astronauts and materials to and from permanently situated moon-based stations, in the deployment and operation of communication satellites, and the assemblage of huge spacecraft in space to carry astronauts to Mars and, in the more distant future, possibly to other planets in the solar system. The space shuttle may prove to be the basis for expanding space frontiers through greater accessibility to outer space by the 1980's and 1990's.

A "sortie can" has been proposed which in conjunction with the space shuttle, may usher in a new era of commercial and scientific exploitation of space resources. It would be a pressurized container capable of accommodating from six to twelve persons and designed to be carried by a space shuttle. When the desired orbit was reached, it could be ejected into its own earth orbit through the hatchway of the space shuttle and retrieved later. If not ejected, its remotely controlled sensoring devices could be exposed or elevated through the hatchway for the purpose of collecting data. In the last two instances, the "sortie can" would remain attached to the space shuttle throughout its stay in earth orbit. The shuttle's payload might consist of research and applications modules which could be placed in earth orbits and flown in formation with the shuttle spacecraft or on their own in gathering space data.

A Space Shuttle Orbiter with a Payload Consisting of a Spacelab Module and a Pallet. Spacelab is a joint venture of the National Aeronautics and Space Administration (NASA) and the European Space Research Organization (ESRO). Such a Shuttle Spaceship could carry a "sortie" as part of its payload. Artist Concept. (Source: NASA - Photo No. 75-HC-5)

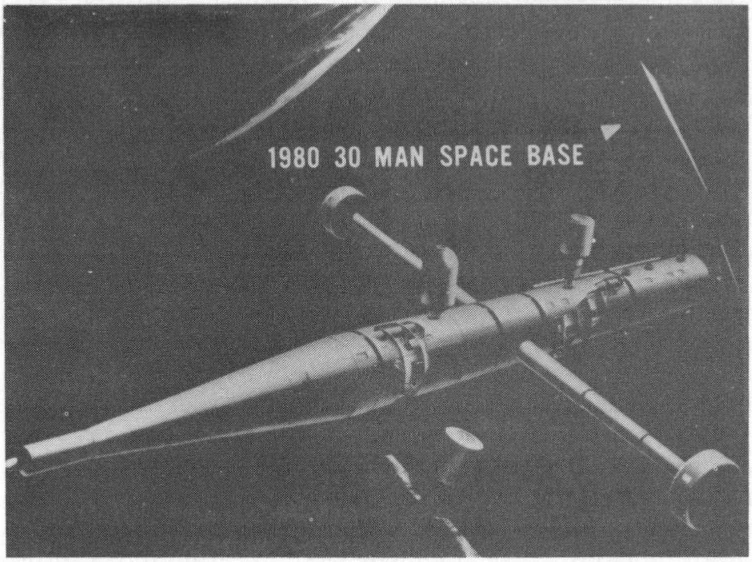

Thirty-Man Space Station, 1980. Artist Concept. (Source: NASA - Photo No. M69-5282)

Emergency Escape Vehicle Proposed for Use to Enable Three Astronauts to Escape from a Space Station (Source: NASA - Photo No. M69-5484)

The space shuttle will require a major investment of time, energy and economic resources, but if successful may have a very significant impact upon the economics of space exploration and exploitation. It has been estimated that, if its potentials are achieved, the cost of putting a pound of material into space may be lowered from two thousand dollars to between one hundred and two hundred dollars.[28] If this reduction is attained, it is likely that new fields will be opened in the use of space benefiting not only the United States but all nations through more adequate surveillance of the earth for survival and safety-rescue activities, more effective inventorying of human and material resources, and more effective communication, making for better understanding and cooperation among the peoples of the world.

The space shuttle will be designed basically to shuttle between earth and earth orbits. It will lift off like a rocket, orbit the earth like a spacecraft, and return to earth like a jet airliner. Its capabilities will simplify the building and development of military, commercial, communication, weather, moon ferry, and planetary probe satellites.

The development of the space shuttle is feasible because: it is technologically possible for it to become operational within six years; it holds promise of reducing the cost of space operations and transforming man's role in space exploration; it will provide the United States with the capability for continuing its scientific investigation and commercial exploitation of space; and it holds promise for the promotion of international cooperation in the use of space.

Without the space shuttle it would not be feasible to develop a large manned space base or space station. Only such an orbiting base can, on a continuing basis, conduct space experiments, try out space manufacturing techniques, and provide a point of departure for deeper manned space missions. Many proposals have been made regarding configurations of space bases or space stations. A fundamental requirement is the building block concept whereby units are transported and assembled in modular fashion in orbit. A 30-man space base might look like the one shown in the figure. Shuttle vehicles would be used to transfer personnel and materials. The space shuttle would also have the means of rescuing personnel in an emergency. However, emergency escape vehicles attached to the space base might serve to bring personnel back to earth in cases where shuttle vehicles could not be provided in time.

REFERENCES

1. *NASA, 18th Semiannual Report to Congress,* July-December, 1967.
2. *NASA, 19th Semiannual Report to Congress,* January-June, 1968.
3. *Post Apollo Space Program, Direction for Future,* Space Group Report to the President, September, 1969—Space Station Module Section.
4. *NASA, 21st Semiannual Report to Congress,* January-June, 1969.
5. It was planned that the project would become operational during the late 1970's. (*NASA, 22nd Semiannual Report to Congress,* July-December, 1969).
6. *NASA, 22nd Semiannual Report to Congress,* July-December, 1969.
7. Captain Charles Conrad, Jr., participated in *Gemini V Flight,* August 21, 1965 (p. 84), *Gemini XI Flight,* September 12, 1966 (p. 88), and *Apollo 12 Flight,* November 14, 1969 (p. 104).
8. Captain Alan L. Bean participated in *Apollo 12 Flight,* November 14, 1969 (p. 104).

9. *Soyuz XI Flight,* June 6, 1971. Cosmonauts Georgi T. Doblovolsky, Vladistov N. Volkov and Viktor I. Palsayev set a space endurance record for a single mission of 23 days, 18 hours, and 22 minutes (p. 138).
10. Footnote 8.
11. The Saturn 1B carrier-rocket and the Apollo spacecraft were brought to the launch pad on August 14, 1973, to be used in a rescue mission to bring the Skylab-3 crew back to earth if the emergency should arise to justify it.
12. Out of the first 25 days in space, more than ninety hours were devoted in gathering information relating to the sun.
13. This solar phenomenon is called a "coronal transient" and takes place two or three times a year.
14. On January 4, Lieutenant Colonel Gerald P. Carr, Lieutenant Colonel R. Pogue, passed Charles Conrad, Jr.'s record of 49 days, 3 hours, 38 minutes and 36 seconds, which he had accumulated in four missions. On completion of the Skylab-4 mission, all members of the Skylab-4 crew surpassed Captain Bean's record of 69 days, 15 hours, 53 minutes, and 59 seconds, set during the Skylab-3 mission.
15. Footnote 11.
16. Statement by Dr. Scheider, Mission Control, Houston, as reported in *The New York Times,* Sunday, December 15, 1973 (24:4).
17. The stabilization system continued to be a problem. On December 7, one of the two remaining gyroscopes had started to function erratically and in the early part of January faltered for two days in a row. It was feared that it would break down completely before the end of the month. However, by January 25, it appeared to show some improvement, raising hopes that it would last until the completion of the mission.
18. There had been some consideration given to a possible revisit to Skylab during the joint Soviet Union and United States earth-orbital flight scheduled for July 15, 1975.
19. On December 1, the astronauts focused the cameras on an intense greenish light blinking upward from earth. The light consisted of beams involved in an experiment to demonstrate the use of laser beams for spacecraft navigation and communication. Lasers are thin beams of intensely amplified light that can travel for thousands of miles.
20. The series of tests involves the technique referred to as "echocardiography," which is the use of sonar waves of extremely high frequency to provide information relating to the internal structure of the heart.
21. *The New York Times,* January 12, 1975 (22:1).
22. Letter from Mr. David W. Garrett, Public Information Officer, Manned Space Flight, National Aeronautics and Space Administration, Washington, September 24, 1975.
23. *NASA, 21st Semiannual Report to Congress,* January to June, 1969, p. 41.
24. *Ibid.*, p. 42.
25. *Space Task Group Report to the President, op. cit.*
26. *NASA, 22nd Semiannual Report to Congress,* July-December, 1969.
27. It has been estimated that it will take at least six years to develop an acceptable space shuttle.
28. M. W. Jack Bell, "The Space Shuttle: Man's First Reusable Space Transport System," *ICAO Bulletin,* April, 1973, p. 14.

Chapter XIII

THE ERA OF INTERNATIONAL COOPERATION IN SPACE EXPLORATION

The Twentieth Century space age was ushered in by the Soviet Union on October 4, 1957 with the successful launching and placing in an earth orbit, the first man-made satellite, Sputnik I. This was a spectacular accomplishment which caught the world by surprise and caused considerable consternation on the part of many space scientists in the United States. This achievement brought to the Soviet Union tremendous prestige as a nation far advanced in space science and technology.

The space age was initiated during a period of adverse political relations between the Soviet Union and the United States involving conflict referred to as the cold war. It was a period of international tension between political ideologies competing for recognition by the developing countries. The circumstances surrounding this rivalry were not conducive to international cooperation in the exploration and use of space.

Leadership in space was considered an indication of superiority bringing respect and prestige to any nation that succeeded in being first in such ventures, and was assumed to have significant propaganda values. Experience revealed that the lack of international space participation and cooperation between the Soviet Union and the United States was more politically oriented than economically or technically oriented.[1] As long as political tensions and mistrusts prevailed between these nations, little international cooperation could be achieved in the exploration and use of space.

A review of the accomplishments in space during the decades (1950-1970) by the Soviet Union and the United States reveals this rivalry. The secretive manner in which the Soviet Union conducted its space activities stands in contrast to the open and informative manner of the United States. The desire for the spectacular and being first in such ventures underscored these accomplishments.

One may speculate as to how much more progress there would have been, had there been the will to cooperate in a free and open manner. It may be argued that this rivalry was necessitated by the international political tensions and distrusts of that period, but it is questionable whether that argument would be as valid today as it was then. A more convincing argument is that the existing international relationships between these two countries are now more conducive to international participation and cooperation in the use of space than at any other time during the space age.

The National Space Policy of the United States

The Act of the Congress of the United States,[2] creating the National Aeronautics and Space Administration (NASA) in 1958 and vesting in it the responsibility for the Nation's civilian space program, set forth a declaration of the nation's space policy and purposes.[3] Incorporated in this declaration were the statements that space activities were to be carried on for peaceful purposes and the benefits derived from them were to be for the good of all mankind. Long-range space studies were to be undertaken for the enhancement of human knowledge and an understanding of the phenomena of the atmosphere and of space and in cooperation with other nations or groups of nations. In support of this declaration, the United States has sought ways and means of implementing its policy objectives through the promotion of international participation and cooperation in the free and open accumulation and dissemination of scientific and technical information relating to the theoretical and applied aspects of the exploration and use of space.

In 1969 the Space Task Group in its report to the President of the United States reiterated the need for international involvement and participation, on a broad basis with respect to space activities. It set forth the following guidelines for achieving the same:

1. The establishment of an international arrangement through which countries may be assured of launch services without being solely and directly dependent upon the United States
2. A division of labor between ourselves and other advanced countries or regional space organizations permitting assumption of primary or joint responsibility for certain scientific or applications tasks in space
3. International sponsorship and support for planetary exploration such as that which was associated with the International Geophysical year.[4]

Changes Making for Increased International Cooperation in Space Activities

In spite of the persistence of international rivalry in space exploration and experimentation, there have occurred fundamental changes in space activities and the political atmosphere in which they are being performed. The political relationships between the Soviet Union and the United States have become more favorable and conducive to international space participation and cooperation. There is apparently now more serious intent to undertake joint projects and exchange scientific and technical space information.

Space exploration has expanded and new challenges are being presented. A new era of planetary investigation has emerged and is demanding greater international participation and cooperation. Planetary exploratory missions have been accomplished with respect to planets of the solar system, including Venus, Mars, Mercury, and Jupiter, and a mission is in progress to fly by Saturn.[5] It is possible that in the not too distant future unmanned missions, involving vast distances, will be undertaken to explore the vicinities of Uranus, Neptune, and Pluto.[6]

Such explorations will be costly, time consuming, and will necessitate increased reliance on highly sophisticated instruments and techniques. Nuclear engines, with enormous thrust, will be required to hurl huge and weighty unmanned spacecraft further and further into the depth of the solar system and beyond into interstellar space. They will be sent on orbital and/or on-the-surface investigatory missions into regions of space where man cannot hope to go now, but where he may eventually go if and when the necessary scientific and technical knowhow becomes available. After an extended period of exploration by unmanned spacecraft, manned orbital and/or on-the-surface missions may be sent to some solar planets.

Space stations, linked by an integrated transportation system and utilizing shuttle spaceships, may assist in opening new frontiers in space. They may be placed in parking orbits of different solar planets and linked with planetary base stations served by space tugs. If such a network comes to pass, it will require international cooperation in order that all mankind may benefit.

Dr. Wernher von Braun, in a highly technical and mathematical study, undertook to explain the scientific feasibility of sending a manned flotilla of spaceships to Mars.[7] He theorized that a flotilla, consisting of ten spaceships, could be assembled in an earth parking orbit and from it dispactched to Mars. The initial stage of the operation would require 46 three-stage ferry spaceships, making 950 flights over a period of about eight months, to transport the parts for the interplanetary spaceships from the earth to the parking orbit. Each ferry flight would consume 5,583 tons of propellants or a total of 5,320,000 tons for the total ferry operation, at an estimated cost of about $500 million for propellants.

Ten spaceships, when fully equipped for the trip to Mars, would have a complement of 70 men. Once the spaceships were assembled and fueled, they would leave the earth-parking orbit and assume a trajectory ultimately putting them in a Martian orbit. Of these ten spaceships, only seven would eventually return to earth. Three of them would be equipped with landing boats in which members of the landing party would soft-land on Mars. Two of the landing boats would return to seven of the orbiting mother spaceships, which would return to the circum-tellurian orbit. The other three unmanned mother spaceships would continue in the circum-Martian orbit.

Dr. von Braun calculated that the return journey to Mars would take more than three and a half years (1,369 days). The breakdown in terms of days would be: transit time from earth to Mars, 260 days; waiting time in circum-Martian orbit, 449 days; stay on Mars (approximately fifty men), 400 days; and transit time from Mars to earth, 260 days. The total payload which could be placed in the circum-Martian orbit would be 600 tons of which 149 tons could be landed on Mars.

Levitt estimated that a roundtrip to Venus would take about two years and one month involving: 145 days to go to Venus; 470 days stayover or waiting time on Venus and 145 days to return to the earth.[8] A visit to Jupiter from earth and return would require 2,207 days or a little over six years broken down as follows: 997 days to go to Jupiter; 213 days for a stopover or waiting time on the planet, and 997 days for the returntrip to earth. A returntrip between earth and Saturn would involve 12 years; between earth and Uranus, 30 years; and between earth and Neptune or Pluto, 61 years. At a velocity of 30 miles per second (108,000 miles per hour) the transit time from earth to Mars would be reduced from 258 to 44 days and from earth to Pluto from 10,972 to 1,916 days. If it were increased to 100 miles per second (360,000 miles per hour), it would still require about ten days transit time to reach Mars and 330 days to reach Pluto.[9]

Many speculative space ventures appear unrealistic and unattainable. Others are more realistic such as the possibility of landing men on Mars. Exploratory unmanned missions have been sent to Venus, Mars, Mercury and Jupiter and other missions are being considered for the exploration of other planets of the solar system.[10]

Opportunities for International Participation and Cooperation in Space and Planetary Exploration.

There are numerous opportunities for international cooperation in research involving earth-orbiting spacecraft and space stations. Such undertakings include the joint investigation of the earth's atmosphere, its magnetic field, and other phenomena. Joint astronomical observations, including studies of solar phenomena, could be carried out from earth and lunar-orbiting satellites or from stations based on the moon and possibly on some far planets in the solar system. In addition, joint on-the-surface exploration of the moon and solar planets could be undertaken with unmanned instrumented space capsules and, in some instances, manned spacecraft.

International cooperation could be sponsored in practical applications. For example, extensive networks of satellites could be coordinated for gathering and disseminating information as a basis for more accurate worldwide forecasting of the weather. Methods for warning nations of impending natural or man-made disasters could be developed jointly through international participation and cooperation. Furthermore, manned spacecraft could be utilized for joint biomedical studies and the coordination of other space experiments. Tracking of spacecraft could provide also opportunities for international joint efforts.

Considerable attention is being given to studies of the earth's resources from earth-orbiting satellites. The location and mapping of these resources could provide ways and means for their more effective utilization. A device was designed to determine the biological productivity of the oceans and bodies of inland water.[11] By detecting the presence of chlorophyll, an ingredient essential to plant life, it could be determined if there were sufficient plant life for the survival of fish in given bodies of water. The device could also be used to detect areas of pollution in inland waters and estuaries where the excessive algae growth make such areas unhabitable for fish life.

Sensing devices could be used for surveying crops in various regions as a basis for projecting the potential food and feed grains supplies. The potential underground water resources could be surveyed from orbiting satellites, which could have an important bearing upon the availability of world water resources[12] and maps of minerals and forests could be designed for worldwide dissemination. In addition, the charting of the expansion of urban areas could be undertaken, showing the extent to which they were encroaching upon the open spaces. Some earth-monitoring satellites[13] could reveal the manner in which desert areas were invading cultivated areas. It is known that the Sahara and Kalahari deserts are slowly encroaching on contiguous agricultural areas, as a result of improper agricultural practices. Joint earth-monitoring missions could assist in developing methods for reclaiming of much of this land.

Examples of International Participation and Cooperation in Space Activities

Beginning with 1970, international space conferences were held dealing with scientific and technical matters, and exchanges took place between space scientists and technicians of the Soviet Union and the United States.

A meeting was held during January, 1970, at which an agreement was concluded covering joint projects and exchanges in connection with near-earth space research investigations of the moon and planets of the solar system, the development of research experimentation in relation to space meteorology, and space applications to the natural environment.

Especially significant were the goodwill tours of the American astronauts to European countries, some of which included the Soviet Union. Colonel Armstrong, the first to set foot upon the surface of the moon, visited the Soviet Union during June 1970, and while there stressed the desirability of closer cooperation between the two countries in the exploration and use of space.

During September 1970, at a meeting of the European Space Conference held in Venice, consideration was given to the role which Europe would play in the development of the space shuttle. Mr. Theo Lefevre, Belgian Minister of Scientific Policy and Planning, reported on his negotiations in Washington. He stated that in return for a European commitment to participate in the development of the space shuttle, the United States would put European spacecraft into orbit until the shuttle was available to perform that function. (The European Space Conference is composed of the Science Minister from the West European countries.) This conference also endorsed a move to upgrade the conference into a counterpart of the National Aeronautics and Space Administration.

Five American space experts of the United States visited Moscow during October 24, 1970 for two days to participate in the first direct talks with the Soviet space experts on the possibility of joint space programs, such as plans for cooperative efforts of manned-orbiting space stations and linking of the spacecraft of the two nations while in an earth orbit. It was concluded to continue the efforts to develop compatible systems of rendezvous and docking but no agreement was reached.

Later (October 29) an agreement was signed in Moscow providing for a series of meetings of three committees of technical experts from each country to draft designs for such equipment and to make plans for a rendezvous and docking of spaceships of the two nations sometime during the late 1970's. *This was the first agreement to be consummated between the Soviet Union and the United States concerning joint efforts in space flights.*[14] It marked the culmination of more than a decade of overtures on the part of the United States for increased joint participation and cooperation in space activities.

During November 1970, there took place at a technical conference held in Moscow an exchange of docking systems designs between the Soviet Union and the United States. This was the first step in the development and implementation of compatible docking systems between the two nations. These cooperative efforts should prove significant in the development of space rescue techniques and other undertakings.

It was announced on May 25, 1971 that Colonel Stafford would be the official representative of the United States at the Belgrade Cosmos International Space Exhibition. While there, he was to participate in the American Day on July 4. The Apollo 14 astronauts were also scheduled to attend the Power Air Show from May 30 to June 6 and visit a number of cities in France.

In August 1971, three joint Soviet and United States space groups met for five days in Moscow, consisting of members of the Soviet Academy of Science and the United States National Aeronautics and Space Administration. This meeting was held in keeping with the results of the Moscow meeting held in January 1970. The groups developed recommendations for joint projects and exchange of space information.

A further indication of the change in attitude on the part of the Soviet Union relating to cooperation in space exploration and other space activities occurred on March 19, 1972, when the Soviet Union, for the first time, permitted an American journalist a one-day visit to its space installations at Zvezdny Gorodok (Star City). At the time of the visit, the Russians were engaged in constructing a complex of buildings to house the new cosmonaut training facilities for manned earth-orbiting laboratories. Mr. John Noble Wilford, of *The New York Times,* was afforded the opportunity of interviewing Major General Vladimir Shatalov, Chief of Cosmonaut Training at the space center and a veteran of three Soyuz space missions,[15] who volunteered a toast for better space cooperation between the Soviet Union and the United States. The fact that a news correspondent was given the opportunity to visit this space center in the Soviet Union was indicative of a change in the Soviet Union's attitude and policy toward international space cooperation.

On April 17, 1972, Mr. Yevgeny Yevhushenko, a guest a Cape Kennedy during the launching of Apollo 16, expressed his admiration for the American space accomplishments and hoped that the two countries would eventually cross the Milky Way together. Further evidence of a shift in the attitude was the recognition given to the Apollo 16 launching in the Soviet press.

Seven Soviet Specialists in aerospace medicine exchanged data with their American counterparts at a meeting in Houston on May 13, 1972.[16] At this meeting, the specialists of the Soviet Union and the United States discussed data gathered by the Soviet cosmonauts during the Soyuz XI and the Salyut 1 missions and those gathered by the American astronauts during the Apollo 16 mission. During their visit to Houston, the Soviet specialists toured the Manned Space Center.

The Agreement Between the United States and the Soviet Union for Cooperation in the Exploration and Use of Outer Space

In Moscow, on May 24, 1972, President Nixon of the United States and Premier Kosygin of the Soviet Union, signed an accord providing for the cooperation between their respective countries in the exploration and use of space.[17] *This agreement represented the first major step in the fulfillment of the kind of relationships in space activities that the United States had been striving to achieve with the Soviet Union for two decades.* This cooperation in the exploration and use of space was to be carried out for peaceful purposes. The results of the scientific research gained from this cooperation were to be for the benefit of the peoples of the United States and of the Soviet Union as well as for all the people of the world. These were lofty ideals but they were in keeping with the declared intent of the space policy of the United States set forth in 1958.[18]

Article 1 of this agreement provided for the development of cooperation in space meteorology, study of the natural environment, exploration of near-earth space, the moon and the planets, and space biology and medicine. The parties agreed to cooperate in taking appropriate measures for the encouragement and achievement of the summary of the results of discussions on space cooperation between the United States' National Aeronautics and Space Administration and the Soviet Union's Academy of Science as of January 21, 1971.

Under Article 2, the parties agreed to promote such cooperation through mutual exchanges of scientific information and delegations, meetings of scientists and specialists of both countries, and other ways as may be agreed to from time to time. It was agreed that joint working groups could be established for the development and implementation of cooperative space programs.

Article 3 dealt with the rendezvous and docking of the spacecraft of the two countries. This was to be accomplished through the development of compatible rendezvous and docking systems for manned spacecraft and space stations, which, when implemented, should enable the two countries to carry out joint scientific experiments and enhance the safety of manned space flights. It was anticipated that the first experimental flight to test the compatible docking mechanisms would take place in 1975, and that an Apollo-type spacecraft would be used by the United States and a Soyuz-type spacecraft by the Soviet Union. The implementation of these projects was to be in accordance with the summary of results of the meetings between the representatives of the National Aeronautics and Space Administration of the United States and those of the Academy of Science of the Union of Soviet Socialist Republics on the development of compatible systems for rendezvous and docking of manned spacecraft and space stations dated April 6, 1972.

The encouragement of international cooperation in dealing with problems in the area of international law relating to the exploration and use of outer space was agreed to under Article 4. The primary purposes of such cooperation were to strengthen and develop legal order in space activities.

Article 5 made provision for the extension, by mutual consent, of the agreement to other areas of cooperative efforts in the exploration and use of space for peaceful purposes.

Finally, Article 6 provided that the agreement would be validated and come into force when the signatures were affixed, and remain in force for five years. Provision was made for its modification and extension by mutual consent of the parties to the accord.

On July 6, 1972, the space officials of the Soviet Union and the United States met for a two-week conference at the Space Center in Houston. The United States' officials proposed that the Apollo spacecraft should be launched 16 to 40 hours before the Soyuz spacecraft because of its greater space capabilities. This was the first meeting between the space officials of the two nations following the signing of an agreement on May 24, by President Richard Nixon and Premier Aleksei N. Kosygin for a joint space flight.

Nine Soviet space engineers arrived in Houston on November 23, 1972, for a ten-day discussion of the planned joint space venture in July 1975. They discussed with their United States counterparts guidance and control communication and tracking methods.

On December 16, 1972, a four-stage United States rocket lifted a West German satellite into a polar orbit of the earth to be used to collect and transmit to West Germany data about the earth's atmosphere. The satellite was placed in an orbit with an apogee of 534 miles and a perigee of 137 miles taking 95 minutes to complete. It was anticipated that it would circle the earth for at least six months.

In Brussels on December 20, 1972, representatives of twenty-one nations attended the European Space Conference. They agreed in principle to participate in the Post-Apollo Program of the United States and to build a European booster rocket. In addition, they decided to unite the European Launchers' Development Organization and the European Space Research Organization into a single agency by January 1, 1974.

The National Aeronautics and Space Administration announced during the later part of January 1973, the names of astronauts who would make up the crew of the

Apollo spacecraft for the joint international manned space flight scheduled for July 15, 1975. This flight was to involve a Soviet manned Soyuz spacecraft and a United States manned Apollo spacecraft. The mission would include a rendezvous and docking of the two spacecraft and exchange visits between the personnel of each spacecraft. Brigadier General Thomas P. Stafford[19] was to be in command of the Apollo phase of the joint mission and would be accompanied by Vance D. Brand, a civilian, as pilot of the Apollo command module and Donald K. Slayton,[20] also a civilian, Director, Flight Crew Operations at the Manned Spacecraft Center in Houston, as pilot of the docking module.

During February 1973, a 12-man delegation from the United States, under the leadership of Mr. Leonard Jaffee of the National Aeronautics and Space Administration met in Moscow with Soviet Space officials of the Institute for Space Research of the Soviet Academy of Science. The purpose of the joint meeting was for an exchange of information in accordance with the Soviet-United States agreement for cooperation in the study and use of space for peaceful ends.

On March 15, 1973, a 39-man delegation, led by Professor Bushuyev, consisting of Soviet engineers and cosmonauts, Major General Vladimir A. Shatalov[21] and Dr. Aleksel S. Yeliseyev,[22] arrived at Johnson Space Center in Houston to participate in the first round of negotiations relating to the joint space mission by the two countries. After about two weeks of discussion, it was agreed that Soviet cosmonauts and United States astronauts would begin joint training in the United States during the coming summer and in the Soviet Union during the autumn. The two nations had previously exchanged drawings of the designs of the spacecraft to be used in the joint flight. The primary matters considered were joint training for the crews and flight controllers, arrangements for communications between Soviet and United States ground control stations, preliminary discussions of scientific experiments, and detailed discussions of flight time-lines. It was also decided that the Apollo spacecraft would be the active spacecraft in effecting the rendezvous and docking with the Soyuz spacecraft because it could carry a greater load of propellants for use in stabilizing the spacecraft during the two days that they would be linked together. It was also agreed that the crews would visit back and forth between the spacecraft by means of a tunnel at the nose of the Apollo spacecraft. At all times, there would be at least one Soviet cosmonaut in the Soyuz spacecraft and one United States astronaut in the Apollo spacecraft. The joint-flight would take place in a 140-mile earth orbit and each spacecraft would be controlled by its own ground stations. The Soyuz spacecraft would have a mission capability of six days and the Apollo spacecraft twelve days.

It was announced during April 1973, that the United States would make available an Application Technology Satellite for use by India in 1975. It was to be used during six hours daily for at least a year to relay direct broadcasts in connection with India's Educational Television Program.

On July 9, 1973, a team of ten Soviet cosmonauts, accompanied by 24 space engineers, arrived in Houston, headed by Professor Konstantin D. Bushuyev and accompanied by Major General Vladimir A. Shatalov. They were there to plan with the United States' space officials the first joint Soviet Union-United States space flight. With the team were Colonel Aleksei A. Leonov, the first man to walk in space and Valery N. Kubasov, a civilian engineer.[23] Colonel Leonov was to be in command of the Soviet Union Soyuz spacecraft in the joint-flight and Mr. Kubasov, flight engineer.

Another example of international space cooperation was demonstrated by 11 European nations which agreed to form the European Space Agency. The purpose of the agency was to participate with the National Aeronautics and Space Administration in a post-Apollo shuttle program to launch the first Western European astronauts into space. It was planned that a European shuttle spacecraft and United States carrier rocket would be launched from Cape Kennedy, an unmanned launcher developed by France would be placed in space using parts from other European countries, and there would be launchings of a series of British communication satellites for ship-to-shore transmission over the important oil routes between Europe and the Persian Gulf. The countries whose ministers attended the conference held in Brussels on August 1, 1973, were Denmark, France, West Germany, The Netherlands, Spain, Switzerland, Great Britain, Italy, Norway, and Sweden. Italy's participation in the cooperative space venture was not confirmed at that time.

The United States signed an agreement on September 24, 1973, with nine Western European nations enabling them to participate in international space research by constructing a space laboratory by 1980. The National Aeronautics and Space Administration was to cooperate through the European Research Organization, representing Great Britain, West Germany, France, Switzerland, the Netherlands, Belgium, Denmark and Spain. The space laboratory would be carried in a United States built shuttle orbiter, and one European astronaut would be expected to take part in the initial mission. The Marshall Space Flight Center was established as the headquarters for the United States' part of the international space laboratory project.

During the early part of October, 1973, when Soyuz XII was in earth orbit, a 45-man National Aeronautics and Space Administration delegation, headed by Mr. Glynn Lunney, visited the Soviet Union. The space technicians were in Moscow to continue technical discussions relating to the joint Soviet Union-United States space flight scheduled for July 1975. The delegation was divided into three groups: the first was concerned with the overall planning of the mission, the second, with radio and cable communications, including television systems, and the third, with the life support systems and the transfer of personnel between the spacecraft while in earth orbit.

On November 18, 1973, six United States astronauts arrived in Moscow for training with their Soviet counterparts in connection with the joint flight set for July 1975. The team consisted of astronauts Thomas P. Stafford and his crew mates, Donald K. Slayton and Vance D. Brand. Accompanying them were three back-up astronauts: Alan L. Bean,[24] Jack R. Lousma[25] and Ronald E. Evans.[26] They were to train jointly with the Soviet cosmonauts at the Soviet Union's space center at Zvezdny Gorodok (Star City).

The sensitivity concerning international trust and understanding was demonstrated when the United States in January 1974 agreed to make available photographs of the Soviet Union's secret space launching complex at Baikonur in Central Asia to the public upon request. It felt that all ERTS satellite information should be placed in the public domain. The photographs were taken by a United States' Earth Resource Observatory satellite orbiting the earth since 1972. The decision raised potential international political issues, especially when international relations with the Soviet Union were somewhat tense, however no problems arose.

During February 1974, the Soviet Union, contrary to its former policy, proposed that the United Nations establish a center for the collection and distribution of data and information obtained by earth-orbiting satellites. Such information would be of great value in detecting mineral resources, ground water supplies, crop and weather forecasting.

Brazil proposed that a treaty be drawn up to restrict space powers from collecting and distributing information obtained by satellites. The treaty would require consent from the governments concerned for remote-sensing activities being conducted over their territories. It would also require that such governments would have full access to data collected and that such information would not be divulged to other countries, organizations, or private commercial interests without their consent.

The United States offered as early as 1973 to supply the United Nations with master copies of all information obtained from its Earth Resources Technology Satellite-1, launched on July 23, 1972. It is still orbiting the earth, fourteen times each day, and transmitting back information. It is estimated that more than a hundred countries have benefited in some manner from the availability of this information.

During April 1974, four Soviet cosmonauts visited Houston to continue preparations for the joint Soviet-United States space mission scheduled for July 15, 1975. United States space personnel visited the Soviet Union for training later.

A further example of international cooperation took place during December 1974, when the United States joined West Germany in placing an 815-pound satellite in an elliptical orbit of the sun. The space capsule blasted off from Cape Canaveral, Florida atop an American Titan-Centaur booster rocket. The West German-built spool-shaped capsule named Helios-1, after the Greek sun god, should pass within 28 million miles of the sun, where the temperature ranges between 370 and 700 degrees Fahrenheit. The spacecraft should complete three orbits of the sun, each requiring 192 days and by March 1976 should begin sending back data relating to the sun's environment. It is anticipated that Helios-2 will be launched during November 1976 and should approach to within 23 million miles of the sun.

During December 1974, it was reported that an agreement between the space scientists of the Soviet Union and the United States had been reached for an unmanned Soviet satellite to carry biological experiments of the United States into an earth orbit for twenty or thirty days during late 1975. The agreement was consummated at a meeting of scientists and engineers of the two nations held in Tashkent, Soviet Union, and was announced on December 11, 1974, by the National Aeronautics and Space Administration. The Soviet Union usually launches a biological research satellite once a year in its Cosmos series. Although the United States has conducted extensive medical experiments in its Skylab missions, it has no current plans for unmanned biological satellites. It will probably not be until the initiation of the Space Shuttle program about 1980 that the United States will undertake to carry out its own biological experiments. This agreement between the two nations represented a major step forward in scientific cooperation in space and will be activated probably after the completion of the Soviet-Union States joint space flight.

The much publicized international joint spaceflight by a Soyuz spacecraft of the Soviet Union and an Apollo spacecraft of the United States was accomplished successfully. It was inaugurated on Tuesday, July 15, 1975 at 8:20 a.m. (EDT) when the Soyuz XIX spacecraft blasted off its pad at the Baikonur Space Center in Soviet Central Asia about fourteen hundred miles southeast of Moscow. Ten minutes later it was in earth orbit with an apogee of 141 miles and a perigee of 121 miles, where it extended its two solar panels for converting the sun's heat into energy. Onboard were Colonel Aleksei A. Leonov, first man to walk in space, commander, and Valery N.

Kubasov, first man to weld metals in space, flight engineer. *The lift-off was televised live, marking the first time this had been permitted by Soviet space officials.* No Western news reporters were allowed at the launching site, only Soviet reporters.

While this was taking place at the Baikonur Space Center, the countdown was in progress at the Kennedy Space Center in Florida, where the Saturn 1B rocket, the most powerful carrier vehicle possessed by the United States since the retirement of the Saturn V, stood waiting for the final signal for lift-off, with the Apollo spacecraft poised on top of it.

At 3:50 p.m. (EDT) on July 15, the massive carrier rocket blasted off, carrying into space Brigadier General Thomas P. Stafford, veteran of three space flights, commander, and two civilians Vance D. Brand, command module pilot, and Donak K. Slayton, docking module pilot, both making their first flights into space. As the spacecraft rose, television cameras, inside the cabin, showed the astronauts as they reacted to the tremendous acceleration. When launched, the Apollo spacecraft was 4,410 miles behind the Soyuz spacecraft in earth orbit and by Wednesday evening, July 16, the gap had been narrowed to about fifteen hundred miles. The initial orbit assumed by the Apollo spacecraft had an apogee of 105.6 miles and a perigee of 95 miles.

Thursday morning July 17, the spacecraft rendezvoused 140 miles above the earth as they passed over Central Europe and at 12:15 p.m. (EDT) docked as they crossed over Western Germany. The Apollo spacecraft moved from below and behind

The Soyuz Spaceship Linked with the Apollo Spaceship in Earth-Orbital Flight, Artist Concept. (Source: NASA - Photo No. 74-HC-336)

the Soyuz spacecraft to a distance of 2,000 feet ahead of it. Then it turned 180 degrees and moved toward the Soyuz spacecraft and eased its docking mechanism into that of the Soyuz at four-tenths of a foot per second. The latches locked and a smooth and firm docking was accomplished with the Soyuz spacecraft.

Before transfers between the crews could take place, Astronauts Stafford and Slayton had to crawl into the docking tunnel sealing the hatch behind them. This was necessary in order to change the atmosphere and pressure of the tunnel to that of the Soyuz spacecraft.

About 3:25 p.m. the hatch at the opposite end of the docking tunnel was opened and Cosmonaut Leonov and Astronaut Stafford crawled toward each other, exchanged greetings, and shook hands. Then Astronauts Stafford and Slayton entered the orbital module of the Soyuz spacecraft where Astronaut Stafford received the Russian "bear hug" from Cosmonaut Leonov. Flags and gifts were exchanged and Cosmonaut Leonov presented Astronaut Stafford with the flag of the United Nations. They received messages from Communist Party Leader Leonid I. Brezhnev and President Gerald Ford of the United States. At the end of the session, lasting about one and a half hours, Astronaut Stafford and Slayton returned to the Apollo spacecraft where Vance Brand had remained at the controls. All of these activities were televised and relayed to earth and viewed on a worldwide network coverage.

Friday, July 18 was a day of transfers between the crews, providing the opportunity for each member of the crew to visit at least once, the spacecraft of the other country. An early morning transfer was made by Astronaut Brand to the Soyuz spacecraft and Cosmonaut Leonov to the Apollo spacecraft. Later, Cosmonaut Leonov and

ASTP CSM/DOCKING MODULE

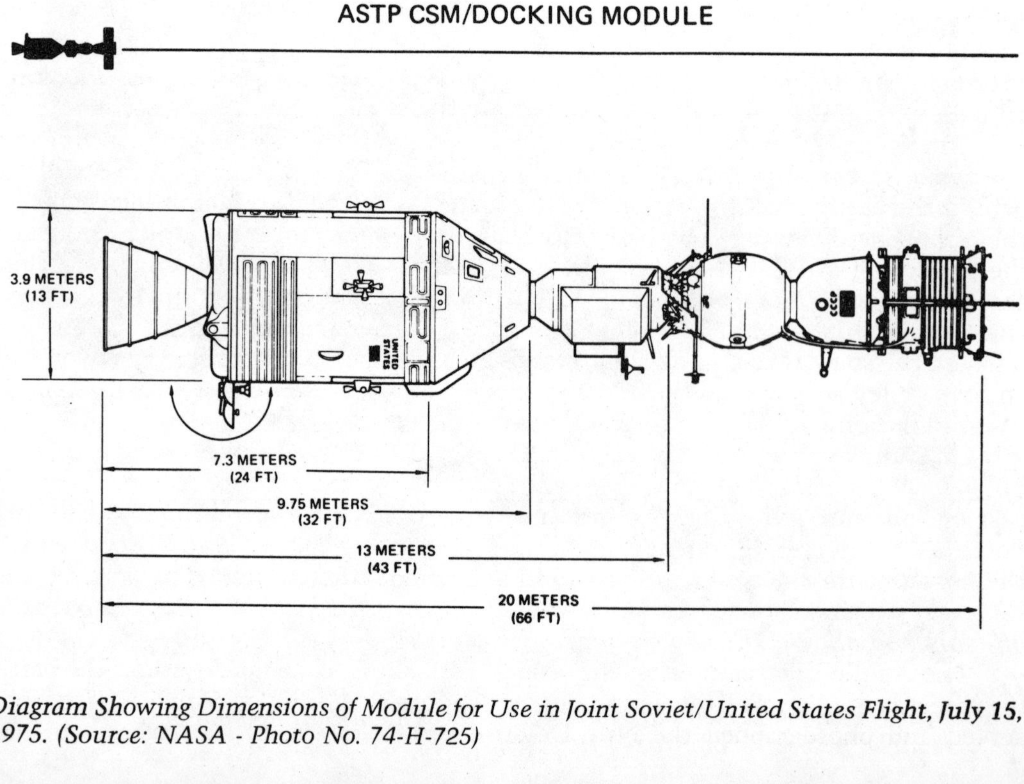

Diagram Showing Dimensions of Module for Use in Joint Soviet/United States Flight, July 15, 1975. (Source: NASA - Photo No. 74-H-725)

Astronaut Stafford transferred to the Soyuz and Astronaut Brand and Cosmonaut Valery Kubasov entered the Apollo. The next transfer involved Astronaut Slayton and Cosmonaut Kubasov who entered the Soyuz. Finally, Astronaut Slayton returned to the Apollo. During the transfers, it was required that there must be at least one member of the crew in the cabin of his own spacecraft.

During the transfers between the crews joint medical and scientific experiments were conducted and the cosmonauts and astronauts shared meals. The cosmonauts spoke in English and the astronauts in Russian. Cosmonaut Leonov presented Astronaut Stafford with a sketch he made of him in the Soyuz cabin during their meeting on the previous day. The two commanders signed a certificate verifying the international flight and joined together a plaque, the two halves of which had been carried into space separately in the two spacecraft. It showed two spacecraft about to dock in space, symbolic of the historic space flight. A farewell ceremony and a news conference was held as the day of transfers came to an end. Then Astronaut Stafford returned to the Apollo cabin, the hatches were sealed, and preparations were made for the next phase of the space flight.

On Saturday, July 19 at 8:02 a.m. (EDT) the spacecraft undocked, while out of contact with ground control stations. When Mission Control reestablished contact with the Apollo crew, the spacecraft was station-keeping 165 feet from the Soyuz spacecraft. It then moved away at about one meter (39.37 inches) per second on a path that blocked the view of the sun from the Soyuz cosmonauts. *This was the first man-made eclipse of the sun* and provided information relating to the sun's corona. The photographs obtained in space of this man-made eclipse are to be compared with those taken of the eclipse by earth-based cameras.

The Apollo spacecraft moved about two hundred meters from the Soyuz spacecraft, stopped, and reversed its direction and moved toward the Soyuz spacecraft which was advancing toward the Apollo spacecraft to dock with it. The docking was successful and the spacecraft remained linked for about three hours before undocking at 11:36 p.m. (EDT).

The Soyuz cosmonauts after undocking spent their last day in space carrying out a series of scientific experiments including photography of the sun as it rose and set behind the earth's horizon to measure the reflection and absorption of solar light by the atmosphere, photography of daytime and dusk horizon of the earth to determine the scattering of light caused by the earth's atmosphere, study of small particles, aerosol, in the air, investigation of the effects of space on growth of microorganisms and the mobility of bacteria, their appearance, genetics and survivability. There was onboard the spacecraft a small aquarium to measure the effects of zero gravity on fish embryo. They also studied the division of cells, their genetic structure, and sensitivity to radiation.

On Monday, July 21 at 6:51 a.m. (EDT) the Soyuz spacecraft touched down 6.2 miles from its landing target near Acholic in Kazakhstan about three hundred miles east of Baikonur Space Center. Within thirty seconds, a helicopter landed carrying doctors, technicians, military personnel and reporters. Soon the hatch was opened and the cosmonauts emerged smiling and in good physical condition. The landing was televised live and shown around the world. *This was the first time that Soviet space officials had allowed this to be done.*

The Apollo spacecraft, after the second undocking, continued to orbit the earth for three days to permit the astronauts to conduct scientific experiments. They observed and photographed the Nile Delta, desert colors in Africa, icebergs in the

Antarctic, cloud features over South America, ocean currents off Argentina and waves off Hawaii. They also observed world vegetation patterns. An interesting scientific search carried on by the astronauts was for a signal from stars that man has never seen. It is in the form of radiation at the extreme end of the ultraviolet part of the electromagnetic spectrum, which is hidden from earth-based telescopes because of absorption by the atmosphere. Some scientists believe it will never be seen; others believe there is a possibility of its being seen in space eventually.

On Thursday, July 24, the Apollo astronauts made a fiery reentry into the earth's atmosphere and at 5:20 p.m. (EDT) splashed down in the Pacific Ocean, 330 miles west of Pearl Harbor. Within forty minutes, the Apollo capsule was hoisted aboard the recovery ship USS *New Orleans* and the astronauts emerged from it tired but in good spirits. They received a telephone message from President Ford congratulating them on their accomplishments in space.

The mission was not without its serious incident. The failure of the astronauts to throw two switches in the final stages of the descent set in motion a series of reactions causing toxic gas to flow into the cabin. Fortunately, General Stafford was able to distribute gas masks and even though Vance Brand passed out, no permanent injury to their lungs resulted. They did have to spend some time under medical observation before being allowed to return to their homes.

The international joint space mission brought to an end the use of the Apollo spacecraft for carrying astronauts into space. It is anticipated that the next astronauts of the United States will travel into space aboard the Space Shuttle, probably by the end of the decade.

It should not be concluded that, because high-level agreements have been signed between nations to cooperate in the exploration and use of space, past rivalries have been set aside completely. In fact, space rivalries and the desire to lead in this field of endeavor will continue to be a strong motivating force in the space relationships between nations. The willingness to cooperate needs to be approached with guarded optimism, especially regarding the Soviet Union and the United States in spite of the favorable responses to overtures on the part of the United States for the peaceful and cooperative exploration and exploitation of space.

Many hurdles, underscored by opposing political attitudes, will have to be overcome before free and open disclosures and exchanges of scientific and technical information relating to space investigations and planetary excursions become a recognized part of international relationships. Nevertheless, the above examples of international cooperation are steps forward when viewed in the light of past international rivalry.

REFERENCES

1. *The Post-Apollo Space Programs, Direction for the Future,* Space Task Group Report to the President, September, 1969.
2. Public Law 85-568, 85th Congress, *National Aeronautics Act of 1958,* July 29, 1958, Section 202(a).
3. *Ibid.,* Section 102 (a), (b), and (c).
4. The Post-Apollo Space Program, *Direction for the Future, op. cit.*
5. *Pioneer 11,* April 5, 1973 (p. 68).
6. *World Book Encyclopedia,* p. 497b:

	Greatest Distance from Earth	Nearest Distance from Earth	Average Distance from Earth
Sun	94,512,000 miles	91,402,000 miles	92,957,000 miles
Moon	251,968 miles	225,742 miles	238,855 miles
Mercury	134,940,000 miles	50,974,000 miles	92,957,000 miles
Venus	161,350,000 miles	24,564,000 miles	92,957,000 miles
Mars	248,890,000 miles	34,389,000 miles	141,640,000 miles
Jupiter	599,960,000 miles	367,320,000 miles	483,640,000 miles
Saturn	1,027,500,000 miles	745,900,000 miles	886,700,000 miles
Uranus	1,959,600,000 miles	1,606,600,000 miles	1,783,100,000 miles
Neptune	2,910,200,000 miles	2,678,000,000 miles	2,794,100,000 miles
Pluto	4,677,100,000 miles	2,669,900,000 miles	3,673,500,000 miles

R. W. Buchheim, *Space Handbook,* (New York: Random House, 1958), p. 32.

Planet	Minimum Launching Velocity (ft./sec.)	Transmit Time
Mercury	44,000	110 days
Venus	38,000	150 days
Mars	38,000	260 days
Jupiter	46,000	2.7 years
Saturn	49,000	6 years
Uranus	51,000	16 years
Neptune	52,000	31 years
Pluto	53,000	46 years

Space Handbook: Astronautics and its Applications, Staff Report of the Select Committee on Astronautics and Space Exploration, U.S. Government Printing Office, 1959, p. 23.

7. Wernher von Braun, *The Mars Project,* (Urbana: The University of Illinois Press, 1953). This volume is based on Das Marsprojekt — A special issue of the magazine Weltraumfahrt, published in Germany, 1952.
8. I. M. Levitt, *Target for Tomorrow,* (New York: Fleet Publishing Corporation, 1959), pp. 242-243.
9. *Ibid.*, p. 244.
10. *Mariner 9*, May 30, 1971 (p. 57); *Pioneer 10,* March 2, 1972 (p. 63); and *Pioneer 11,* April 5, 1973 (p. 68).
11. This device was described by Mr. John C. Arversen of the National Aeronautics and Space Administration, Ames Research Center, at the International Aeronautics Congress held in Brussels, September 1971. (NYT, S-21, 46:1)
12. Mr. William S. Fischer of the United States Geological Survey at the International Aeronautical Congress held in Brussels, September 1971, presented an example of how underground water could be located from space, citing the discovery of photographs taken from an earth-orbiting spacecraft of unknown faults cutting across the Southern states. They revealed how water had been leaking from a reservoir and indicated to well diggers where they would most likely find water. (NYT, S-21, 46:1)

13. Benzt F. Lundholm, Swedish Natural Science Research Council (NYT, S-21, 46:1).
14. An agreement was signed during the mid-sixties providing for the exchange of meteorological information obtained by weather satellites, but its implementation has been sporadic. The Russians and Americans have taken part jointly in satellite communication experiments.
15. Major General Vladimir A. Shatalov participated in *Soyuz IV Flight*, January 14, 1969 (p. 18); *Soyuz VIII Flight*, October 13, 1969 (p. 19); and *Soyuz X Flight*, April 23, 1971 (p. 136).
16. This was the second meeting between the groups, the first having taken place in Moscow, during October 1971.
17. The text of the United States-Soviet Agreement on Cooperation in Space Exploration as distributed by The Associated Press and cited in *The New York Times*, May 25, 1972, p. 14:1.
18. Footnote 2.
19. *Gemini VI-A Flight*, December 15, 1965 (p. 85); *Gemini IX-A Flight*, June 3, 1966 (p. 86); and *Apollo 10 Flight*, May 18, 1969 (p. 97).
20. One of the original Mercury astronauts selected in April 1959, but was grounded in 1962, because of a minor irregularity in heart rhythm and was not returned to astronaut-flight status until the spring of 1972.
21. Footnote 15.
22. Aleksei S. Yeliseyev, participated in *Soyuz V Flight*, January 15, 1969 (p. 18); *Soyuz VIII Flight*, October 13, 1969 (p. 19); and *Soyuz X Flight*, April 23, 1971 (p. 136).
23. Colonel Leonov was a member of *Voshkod Flight II*, March 18, 1965 (p. 13), and Mr. Kubasov was a member of *Soyuz Flight VI*, October 11, 1969 (p. 19).
24. *Apollo 12 Flight*, November 14, 1969 (p. 104), and *Skylab 3 Mission*, July 28, 1973 (p. 156).
25. *Skylab 3 Mission*, July 28, 1973 (p. 156).
26. *Apollo 17 Flight*, December 7, 1972 (p. 127).

EPILOGUE

"When man has conquered all the depths of space, and all the mysteries of time, then he will be but still beginning!"

H. G. Wells (The Shape of Things to Come)[1]

Perhaps in the distant future a great Noah's Ark of Space will be built to journey to some distant star or stars.[2] The ark would have to be large enough to accommodate an entire civilization — an artificial world in transit. In the present state of the art much of it would have to be constructed out of materials and resources found on the moon and would probably weigh over a million tons. Housed in it, one generation would set out for some distant star which later generations would finally reach.

If such a spaceship could be constructed and if whole civilizations with advanced institutions and techniques could live and maintain themselves while in transit, it would take ages for it to traverse the vast interstellar distances. If it were capable of traveling ten miles per second, it could only cover a third of a billion miles a year.

Consider, therefore, how long it would take to travel between earth and the following stars:[3] Proxima Centauri, the nearest star to earth, except the sun, 4.27 light-years away; Alpha Centauri, the nearest bright star, except the sun, 4.31 light-years; Sirius, the brightest star, 8.73 light-years; and the Magellanic Cloud, the nearest galaxy to earth, 137,000 light-years. The maximum speed attainable is considered to be close to the speed of light.

So vast are interstellar distances that it is unlikely man himself, using the technology with which he is familiar, could traverse them. Even though manned interstellar flights may be beyond man's current or foreseeable technology, he will no doubt succeed in sending electronic signals to distant stars. But it is quite another matter to establish coherent communications with intelligent beings, assuming that they may inhabit some of them, because of the elapsed time required for electronic impulses to traverse the enormous distances which separate the distant stars from earth. The enormity of the problem becomes more evident when it is realized that the light from distant stars takes incomprehensible periods of time to reach earth. Of the billions of stars in the Universe, only seventeen of them are within twelve light-years of it, and some are believed to be as far away as one thousand six hundred billion light-years.

Man has already traveled in person to and from the moon and landed upon its surface. He has explored by means of unmanned instrumented space capsules the planets of Venus, Mercury, Mars, and Jupiter and has sent unmanned spacecraft to fly by Saturn and to go beyond into outer space.

In the relatively near future he may, in person, visit some of the planets of the solar system. In generations to come he may even succeed in establishing communications with other intelligent beings, if such there be on other planets, and uncover the secrets of the origin and purpose of the Universe and even Man himself. It is difficult to conclude, based upon our limited understanding of Man's potential, that such things are impossible:

> The road to infinity, a treadmill to oblivion for the early dreamers, is now a pathway to the Stars. Groping upward from his trackless primeval beginnings, Man looked up to the phantasmagoric array of stars and planets and told himself. One Fine Day . . . [4]

REFERENCES

1. Cited in Kenneth W. Gatland, (Ed.), *Project Satellite,* (London: Allan Wingate, 1958), Chapter 4, "Interplanetary Flight," by A. V. Cleaver, p. 134.
2. I. M. Levitt, *Target for Tomorrow,* (New York: Fleet Publishing Corporation, 1959), pp. 270-274.
3. To measure interstellar distances, the astronomers often use the distance between the earth and the moon as an astronomical unit and express the distances between heavenly bodies as being so many times that unit. The more appropriate measuring unit is the "light-year," the distance which light travels in one year. The speed of light is approximately 186,000 miles per second and therefore travels six million million miles (6,000,000,000,000 miles) in one year. (*The World Book Encyclopedia,* p. 4440)
4. Jules Bergman, *Ninety-Seconds to Space, The X-15 Story,* (Garden City, New York: Doubleday and Company, 1960), p. 211.

BIBLIOGRAPHY

Adams, Carsbie C., *Space Flight, Satellites, Spaceships, Space Stations and Space Travel Explained,* (New York: McGraw-Hill Book Company, Inc., 1958).

Adler, Irving, *Seeing the Earth From Space, What the Man-Made Moons Tell Us,* (New York: The John Day Company, 1961).

Armstrong, Neil, *et al., First on the Moon,* (Boston: Little, Brown and Company, 1970).

Beard, R. B. and Rotherham, A. C., *Space Flight and Satellite Vehicles,* (New York: Pitman Publishing Corporation, 1960).

Beeland, Lee and Wells, Robert, *Space Satellite, The Story of the Man-Made Moon,* (Englewood Cliffs, N.J.: Prentice-Hall, Inc., 1957).

Bergaust, Erik and Beller, William, *Satellites!* (Garden City, N.Y.: Hanover House, 1956).

Bergman, Jules, *Ninety Seconds to Space, The X-15 Story,* (Garden City, N.Y.: Doubleday & Company, 1960).

Bonestell, Chesley and Ley, Willy, *Conquest of Space,* (New York: The Viking Press, 1949).

Branley, Franklyn M., *Exploration of the Moon,* (Garden City, N.J.: The National History Press, Published for the American Museum of Natural History, 1964).

Buchheim, Robert W., *Space Handbook: Aeronautics and Its Applications,* (New York: Random House, 1959).

Caidin, Martin, *Countdown For Tomorrow, The Inside Story of Earth Satellites, Rockets and Missiles and the Race Between American and Soviet Science.* (New York: E. P. Dutton & Company, Inc., 1958).

———, *Vanguard, The Story of the First Man-made Satellite,* (New York: E.P. Dutton & Company, Inc., 1957).

———, *War For the Moon,* (New York: E. P. Dutton & Company, Inc., 1959).

Clarke, Arthur C., *Interplanetary Flight, An Introduction to Astronautics,* (New York: Harper & Brothers, Publishers, 1951).

———, *The Making of A Moon, The Story of the Satellite Program.* (New York: Harper & Brothers, Publishers, 1958).

Cox, Donald and Stoiko, Michael, *Space Power, What It Means to You,* (Philadelphia: The John C. Winston Company, 1958).

De La Ferte, Philip, Joubert (Sir) *Rocket,* (New York: Philosophical Society, 1957).

Del Rey, Lester, *Rockets Through Space, The Story of Man's Preparations to Explore the Universe,* (Philadelphia: The John C. Winston Company, 1957).

DuBridge, Lee A., *Introduction to Space,* (New York: Columbia University Press, 1960).

Emme, Eugene, *Aeronautics and Astronautics, An American Chronology of Science and Technology in Exploration of Space, 1915-1960,* (Washington: National Aeronautics and Space Administration, 1961).

Evans, F. T. and Howard, H. D., *Outlook on Space,* (London: George Allen & Unwin Ltd., 1965).

Gatland, Kenneth W., *Telecommunication Satellites.* (London: Iliffe Books—Englewood Cliffs, N.J.: Prentice-Hall, Inc., 1964).

———, *Project Satellite,* (London: Allan Wingate, 1958).

Gilzin, Karl, (Translated by Pauline Rose), *Sputniks and After,* (London: Macdonald, 1959).
Godwin, Felix, *The Exploration of the Solar System,* (New York: Plenum Press, Inc., 1960).
Goodwin, Harold Leland, *Space, Frontier Unlimited,* (New York: D. Van Nostrand Company, Inc., 1962).
Groves, G. V., *Space Travel, The World Around Us,* (London: The Educational Supply Association, 1959).
Haley, Andrew G., *Rocketry and Space Exploration,* (New York: D. Van Nostrand Company, Inc., 1958).
Hirch, Lester M., (Ed.) *Man and Space,* (New York: Pitman Publishing Corporation, 1961).
Holmes, David C., *What's Going on In Space, A Chronicle of Man's Exploration Into Space Beyond This Earth,* (New York: Funk & Wagnalls Company, 1958).
Jung, C. G., (Translated by R. F. C. Hull), *Flying Saucers, A Modern Myth of Things Seen in the Skies,* (New York: Harcourt, Brace and Company, 1959).
Keyhoe, Donald E., (Major), *Flying Saucers from Outer Space,* (New York: Henry Holt and Company, 1953).
Krieger, F. J., *Behind the Sputniks, A Survey of Soviet Space Science,* (Washington: Public Affairs Press, 1958).
Lapp, Ralph E., *Man and Space, The Next Decade,* (New York: Harper & Brothers, 1961).
———, *Space Science,* (New York: American Library Association in Cooperation with the Public Affairs, Inc., 1962).
Leonard, Jonathan Norton, *Flight into Space, The Facts, Fancies and Philosophy,* (New York: Random House, Inc., 1953).
Leslie, Desmond and Adamski, George, *Flying Saucers Have Landed,* (London: Werner Laurie, 1953).
Levitt, I. M., *Target For Tomorrow, Space Travel of the Future,* (New York: Fleet Publishing Corporation, 1959).
Ley, Willy, *Harnessing Space,* (New York: The Macmillan Company, 1963).
———, *Rockets, Missiles, and Space Travel,* (New York: The Viking Press, 1957).
Mallan, Lloyd, *Men, Rockets and Space Rats,* (New York: Julian Messner Ltd., 1955).
Menzel, Donald and Boyd, Lyle G., *The World of Flying Saucers, A Scientific Examination of a Major Myth of the Space Age,* (Garden City, N.Y.: Doubleday & Company, Inc., 1963).
Moore, Patrick, *Rockets and Earth Satellites,* (London: Frederick Muller Ltd., 1962).
Oberth, Hermann, *Man into Space, New Projects For Rocket and Space Travel,* (New York: Harper & Brothers Publishers, 1957).
Odishaw, Hugh, (Ed.), *The Challenges of Space,* (Chicago: University of Chicago Press, 1962).
———, *Outer Space, Prospects for Man and Society,* Englewood Cliffs: Prentice-Hall, Inc., 1962. For the American Assembly Columbia University.
Richardson, Robert S., (Ed.), *Man and the Moon,* (Cleveland: The World Publishing Company, 1961).
Ryan, Cornelius (ed.), *Across the Space Frontier,* (New York: The Viking Press, 1953).
———, (ed.), *Conquest of the Moon,* (New York: The Viking Press, 1953).
Scully, Frank, *Behind the Flying Saucers,* (New York: Henry Holt and Company, 1950).
Shirley, Thomas, *Men of Space,* (New York: Clinton Company, 1960).
———, *Soviet Writings on Earth Satellites and Space,* (New York: The Citadel Press, 1958).
Sullivan, Walter, *America's Race for the Moon, The New York Times Story of Project Apollo,* (New York: Random House, 1962).
Vasil'ev, M., *Sputnik into Space,* (New York: The Dial Press, 1958).
von Braun, Wernher, *First Men to the Moon,* (New York: Holt, Rinehart and Winston, 1960).
———, *The Mars Project,* (Urbana: The University of Illinois Press, 1953).
Williams, Beryl and Epstein, Samuel, *The Rocket Pioneers, On the Road to Space,* (New York:

Julian Messner, Inc., 1958).

Witkin, Richard, (ed.), *The Challenge of the Sputniks,* (New York: Doubleday & Company, 1958).

Woodbury, David O., *Around the World in 90 Minutes, The Fabulous True Story of the Man-Made Moons* (New York: Harcourt, Brace and Company, 1958).

INDEX

A

B